Graph Algorithms
Practical Examples in Apache Spark and Neo4j

Mark Needham and Amy E. Hodler

Beijing · Boston · Farnham · Sebastopol · Tokyo

Graph Algorithms

by Mark Needham and Amy E. Hodler

Published by O'Reilly Media, Inc., 1005 Gravenstein Highway North, Sebastopol, CA 95472.

O'Reilly books may be purchased for educational, business, or sales promotional use. Online editions are also available for most titles (*http://oreilly.com*). For more information, contact our corporate/institutional sales department: 800-998-9938 or *corporate@oreilly.com*.

Acquisitions Editor: Jonathan Hassell	**Indexer:** Judy McConville
Development Editor: Jeff Bleiel	**Interior Designer:** David Futato
Production Editor: Deborah Baker	**Cover Designer:** Karen Montgomery
Copyeditor: Tracy Brown	**Illustrator:** Rebecca Demarest
Proofreader: Rachel Head	

May 2019:   First Edition

Revision History for the First Edition

2019-04-15:	First Release
2019-05-16:	Second Release
2020-06-05:	Third Release
2021-04-23:	Fourth Release

See *http://oreilly.com/catalog/errata.csp?isbn=9781492047681* for release details.

The O'Reilly logo is a registered trademark of O'Reilly Media, Inc. *Graph Algorithms*, the cover image of a European garden spider, and related trade dress are trademarks of O'Reilly Media, Inc.

This work is part of a collaboration between O'Reilly and Neo4j. See our statement of editorial independence (*http://www.oreilly.com/about/editorial_independence.html*).

978-1-492-05781-9

[LSI]

Table of Contents

Preface

The world is driven by connections—from financial and communication systems to social and biological processes. Revealing the meaning behind these connections drives breakthroughs across industries in areas such as identifying fraud rings and optimizing recommendations to evaluating the strength of a group and predicting cascading failures.

As connectedness continues to accelerate, it's not surprising that interest in graph algorithms has exploded because they are based on mathematics explicitly developed to gain insights from the relationships between data. Graph analytics can uncover the workings of intricate systems and networks at massive scales—for any organization.

We are passionate about the utility and importance of graph analytics as well as the joy of uncovering the inner workings of complex scenarios. Until recently, adopting graph analytics required significant expertise and determination, because tools and integrations were difficult and few knew how to apply graph algorithms to their quandaries. It is our goal to help change this. We wrote this book to help organizations better leverage graph analytics so that they can make new discoveries and develop intelligent solutions faster.

What's in This Book

This book is a practical guide to getting started with graph algorithms for developers and data scientists who have experience using Apache Spark™ or Neo4j. Although our algorithm examples utilize the Spark and Neo4j platforms, this book will also be helpful for understanding more general graph concepts, regardless of your choice of graph technologies.

The first two chapters provide an introduction to graph analytics, algorithms, and theory. The third chapter briefly covers the platforms used in this book before we dive into three chapters focusing on classic graph algorithms: pathfinding, centrality, and community detection. We wrap up the book with two chapters showing how graph algorithms are used within workflows: one for general analysis and one for machine learning.

At the beginning of each category of algorithms, there is a reference table to help you quickly jump to the relevant algorithm. For each algorithm, you'll find:

- An explanation of what the algorithm does
- Use cases for the algorithm and references to where you can learn more
- Example code providing concrete ways to use the algorithm in Spark, Neo4j, or both

Conventions Used in This Book

The following typographical conventions are used in this book:

Italic
> Indicates new terms, URLs, email addresses, filenames, and file extensions.

`Constant width`
> Used for program listings, as well as within paragraphs to refer to program elements such as variable or function names, databases, data types, environment variables, statements, and keywords.

`Constant width bold`
> Shows commands or other text that should be typed literally by the user.

`Constant width italic`
> Shows text that should be replaced with user-supplied values or by values determined by context.

> This element signifies a tip or suggestion.

> This element signifies a general note.

This element indicates a warning or caution.

Using Code Examples

Supplemental material (code examples, exercises, etc.) is available for download at *https://bit.ly/2FPgGVV*.

This book is here to help you get your job done. In general, if example code is offered with this book, you may use it in your programs and documentation. You do not need to contact us for permission unless you're reproducing a significant portion of the code. For example, writing a program that uses several chunks of code from this book does not require permission. Selling or distributing examples from O'Reilly books does require permission. Answering a question by citing this book and quoting example code does not require permission. Incorporating a significant amount of example code from this book into your product's documentation does require permission.

We appreciate, but generally do not require, attribution. An attribution usually includes the title, author, publisher, and ISBN. For example: "*Graph Algorithms* by Amy E. Hodler and Mark Needham (O'Reilly). Copyright 2019 Amy E. Hodler and Mark Needham, 978-1-492-05781-9."

If you feel your use of code examples falls outside fair use or the permission given above, feel free to contact us at *permissions@oreilly.com*.

O'Reilly Online Learning

 For more than 40 years, O'Reilly has provided technology and business training, knowledge, and insight to help companies succeed.

Our unique network of experts and innovators share their knowledge and expertise through books, articles, and our online learning platform. O'Reilly's online learning platform gives you on-demand access to live training courses, in-depth learning paths, interactive coding environments, and a vast collection of text and video from O'Reilly and 200+ other publishers. For more information, please visit *http://oreilly.com*.

How to Contact Us

Please address comments and questions concerning this book to the publisher:

O'Reilly Media, Inc.
1005 Gravenstein Highway North
Sebastopol, CA 95472
800-998-9938 (in the United States or Canada)
707-829-0515 (international or local)
707-829-0104 (fax)

We have a web page for this book, where we list errata, examples, and any additional information. You can access this page at *https://oreil.ly/graph-algorithms*.

To comment or ask technical questions about this book, send email to *bookquestions@oreilly.com*.

For news and more information about our books and courses, see our website at *http://www.oreilly.com*.

Find us on Facebook: *http://facebook.com/oreilly*

Follow us on Twitter: *http://twitter.com/oreillymedia*

Watch us on YouTube: *http://www.youtube.com/oreillymedia*

Acknowledgments

We've thoroughly enjoyed putting together the material for this book and thank all those who assisted. We'd especially like to thank Michael Hunger for his guidance, Jim Webber for his invaluable edits, and Tomaz Bratanic for his keen research. Finally, we greatly appreciate Yelp permitting us to use its rich dataset for powerful examples.

Foreword

What do the following things all have in common: marketing attribution analysis, anti-money laundering (AML) analysis, customer journey modeling, safety incident causal factor analysis, literature-based discovery, fraud network detection, internet search node analysis, map application creation, disease cluster analysis, and analyzing the performance of a William Shakespeare play. As you might have guessed, what these all have in common is the use of graphs, proving that Shakespeare was right when he declared, "All the world's a graph!"

Okay, the Bard of Avon did not actually write *graph* in that sentence, he wrote *stage*. However, notice that the examples listed above all involve entities and the relationships between them, including both direct and indirect (transitive) relationships. Entities are the nodes in the graph—these can be people, events, objects, concepts, or places. The relationships between the nodes are the edges in the graph. Therefore, isn't the very essence of a Shakespearean play the active portrayal of entities (the nodes) and their relationships (the edges)? Consequently, maybe Shakespeare could have written *graph* in his famous declaration.

What makes graph algorithms and graph databases so interesting and powerful isn't the simple relationship between two entities, with A being related to B. After all, the standard relational model of databases instantiated these types of relationships in its foundation decades ago, in the entity relationship diagram (ERD). What makes graphs so remarkably important are directional relationships and transitive relationships. In directional relationships, A may cause B, but not the opposite. In transitive relationships, A can be directly related to B and B can be directly related to C, while A is not directly related to C, so that consequently A is transitively related to C.

With these transitivity relationships—particularly when they are numerous and diverse, with many possible relationship/network patterns and degrees of separation between the entities—the graph model uncovers relationships between entities that otherwise may seem disconnected or unrelated, and are undetected by a relational database. Hence, the graph model can be applied productively and effectively in many network analysis use cases.

Consider this marketing attribution use case: person A sees the marketing campaign; person A talks about it on social media; person B is connected to person A and sees the comment; and, subsequently, person B buys the product. From the marketing campaign manager's perspective, the standard relational model fails to identify the attribution, since B did not see the campaign and A did not respond to the campaign. The campaign looks like a failure, but its actual success (and positive ROI) is discovered by the graph analytics algorithm through the transitive relationship between the marketing campaign and the final customer purchase, through an intermediary (entity in the middle).

Next, consider an anti-money laundering (AML) analysis case: persons A and C are suspected of illicit trafficking. Any interaction between the two (e.g., a financial transaction in a financial database) would be flagged by the authorities, and heavily scrutinized. However, if A and C never transact business together, but instead conduct financial dealings through safe, respected, and unflagged financial authority B, what could pick up on the transaction? The graph analytics algorithm! The graph engine would discover the transitive relationship between A and C through intermediary B.

In internet searches, major search engines use a hyperlinked network (graph-based) algorithm to find the central authoritative node across the entire internet for any given set of search words. The directionality of the edge is vital in this case, since the authoritative node in the network is the one that many other nodes point at.

With literature-based discovery (LBD)—a knowledge network (graph-based) application enabling significant discoveries across the knowledge base of thousands (or even millions) of research journal articles—"hidden knowledge" is discovered only through the connection between published research results that may have many degrees of separation (transitive relationships) between them. LBD is being applied to cancer research studies, where the massive semantic medical knowledge base of symptoms, diagnoses, treatments, drug interactions, genetic markers, short-term results, and long-term consequences could be "hiding" previously unknown cures or beneficial treatments for the most impenetrable cases. The knowledge could already be in the network, but we need to connect the dots to find it.

Similar descriptions of the power of graphing can be given for the other use cases listed earlier, all examples of network analysis through graph algorithms. Each case deeply involves entities (people, objects, events, actions, concepts, and places) and their relationships (touch points, both causal and simple associations).

When considering the power of graphing, we should keep in mind that perhaps the most powerful node in a graph model for real-world use cases might be "context." Context may include time, location, related events, nearby entities, and more.

Incorporating context into the graph (as nodes and as edges) can thus yield impressive predictive analytics and prescriptive analytics capabilities.

Mark Needham and Amy Hodler's *Graph Algorithms* aims to broaden our knowledge and capabilities around these important types of graph analyses, including algorithms, concepts, and practical machine learning applications of the algorithms. From basic concepts to fundamental algorithms to processing platforms and practical use cases, the authors have compiled an instructive and illustrative guide to the wonderful world of graphs.

— Kirk Borne, PhD
Principal Data Scientist and Executive Advisor
Booz Allen Hamilton
March 2019

CHAPTER 1

Introduction

Graphs are one of the unifying themes of computer science—an abstract representation that describes the organization of transportation systems, human interactions, and telecommunication networks. That so many different structures can be modeled using a single formalism is a source of great power to the educated programmer.

—*The Algorithm Design Manual*, by Steven S. Skiena (Springer),
Distinguished Teaching Professor of Computer Science at Stony Brook University

Today's most pressing data challenges center around relationships, not just tabulating discrete data. Graph technologies and analytics provide powerful tools for connected data that are used in research, social initiatives, and business solutions such as:

- Modeling dynamic environments from financial markets to IT services
- Forecasting the spread of epidemics as well as rippling service delays and outages
- Finding predictive features for machine learning to combat financial crimes
- Uncovering patterns for personalized experiences and recommendations

As data becomes increasingly interconnected and systems increasingly sophisticated, it's essential to make use of the rich and evolving relationships within our data.

This chapter provides an introduction to graph analysis and graph algorithms. We'll start with a brief refresher about the origin of graphs before introducing graph algorithms and explaining the difference between graph databases and graph processing. We'll explore the nature of modern data itself, and how the information contained in connections is far more sophisticated than what we can uncover with basic statistical methods. The chapter will conclude with a look at use cases where graph algorithms can be employed.

What Are Graphs?

Graphs have a history dating back to 1736, when Leonhard Euler solved the "Seven Bridges of Königsberg" problem. The problem asked whether it was possible to visit all four areas of a city connected by seven bridges, while only crossing each bridge once. It wasn't.

With the insight that only the connections themselves were relevant, Euler set the groundwork for graph theory and its mathematics. Figure 1-1 depicts Euler's progression with one of his original sketches, from the paper "Solutio problematis ad geometriam situs pertinentis" (*http://bit.ly/2TV6sgx*).

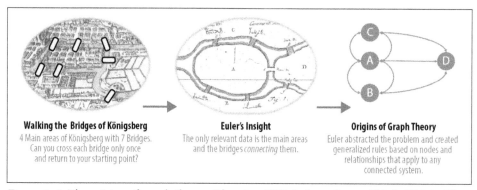

Walking the Bridges of Königsberg
4 Main areas of Königsberg with 7 Bridges. Can you cross each bridge only once and return to your starting point?

Euler's Insight
The only relevant data is the main areas and the bridges *connecting* them.

Origins of Graph Theory
Euler abstracted the problem and created generalized rules based on nodes and relationships that apply to any connected system.

Figure 1-1. The origins of graph theory. The city of Königsberg (https://bit.ly/2JCyLvB) included two large islands connected to each other and the two mainland portions of the city by seven bridges. The puzzle was to create a walk through the city, crossing each bridge once and only once.

While graphs originated in mathematics, they are also a pragmatic and high fidelity way of modeling and analyzing data. The objects that make up a graph are called nodes or vertices and the links between them are known as relationships, links, or edges. We use the terms *nodes* and *relationships* in this book: you can think of nodes as the nouns in sentences, and relationships as verbs giving context to the nodes. To avoid any confusion, the graphs we talk about in this book have nothing to do with graphing equations or charts as in Figure 1-2.

Looking at the person graph in Figure 1-2, we can easily construct several sentences which describe it. For example, person A lives with person B who owns a car, and person A drives a car that person B owns. This modeling approach is compelling because it maps easily to the real world and is very "whiteboard friendly." This helps align data modeling and analysis.

But modeling graphs is only half the story. We might also want to process them to reveal insight that isn't immediately obvious. This is the domain of graph algorithms.

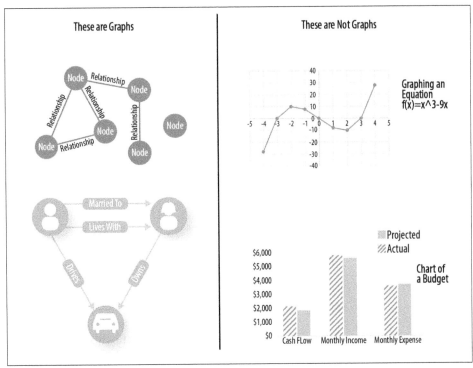

These are Graphs

These are Not Graphs

Graphing an
Equation
f(x)=x^3-9x

Projected
Actual

Chart of
a Budget

Figure 1-2. A graph is a representation of a network, often illustrated with circles to represent entities which we call nodes, and lines to represent relationships.

What Are Graph Analytics and Algorithms?

Graph algorithms are a subset of tools for graph analytics. Graph analytics is something we do—it's the use of any graph-based approach to analyze connected data. There are various methods we could use: we might query the graph data, use basic statistics, visually explore the graphs, or incorporate graphs into our machine learning tasks. Graph pattern–based querying is often used for local data analysis, whereas graph computational algorithms usually refer to more global and iterative analysis. Although there is overlap in how these types of analysis can be employed, we use the term *graph algorithms* to refer to the latter, more computational analytics and data science uses.

Network Science

Network science is an academic field strongly rooted in graph theory that is concerned with mathematical models of the relationships between objects. Network scientists rely on graph algorithms and database management systems because of the size, connectedness, and complexity of their data.

There are many fantastic resources for complexity and network science. Here are a few references for you to explore.

- *Network Science* (*http://networksciencebook.com/*), by Albert-László Barabási, is an introductory ebook
- Complexity Explorer (*https://www.complexityexplorer.org/*) offers online courses
- The New England Complex Systems Institute (*http://necsi.edu/*) provides various resources and papers

Graph algorithms provide one of the most potent approaches to analyzing connected data because their mathematical calculations are specifically built to operate on relationships. They describe steps to be taken to process a graph to discover its general qualities or specific quantities. Based on the mathematics of graph theory, graph algorithms use the relationships between nodes to infer the organization and dynamics of complex systems. Network scientists use these algorithms to uncover hidden information, test hypotheses, and make predictions about behavior.

Graph algorithms have widespread potential, from preventing fraud and optimizing call routing to predicting the spread of the flu. For instance, we might want to score particular nodes that could correspond to overload conditions in a power system. Or we might like to discover groupings in the graph which correspond to congestion in a transport system.

In fact, in 2010 US air travel systems experienced two serious events involving multiple congested airports that were later studied using graph analytics. Network scientists P. Fleurquin, J. J. Ramasco, and V. M. Eguíluz used graph algorithms to confirm the events as part of systematic cascading delays and use this information for corrective advice, as described in their paper, "Systemic Delay Propagation in the US Airport Network" (*https://www.nature.com/articles/srep01159/*).

To visualize the network underpinning air transportation Figure 1-3 was created by Martin Grandjean for his article, "Connected World: Untangling the Air Traffic Network" (*http://bit.ly/2CDdDiR*). This illustration clearly shows the highly connected structure of air transportation clusters. Many transportation systems exhibit a concentrated distribution of links with clear hub-and-spoke patterns that influence delays.

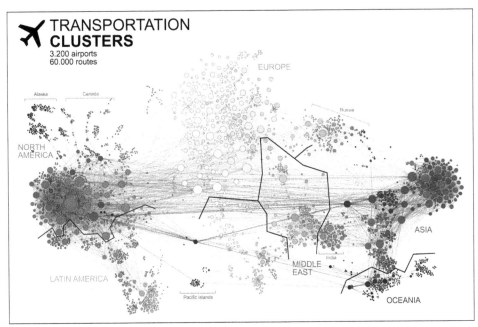

Figure 1-3. Air transportation networks illustrate hub-and-spoke structures that evolve over multiple scales. These structures contribute to how travel flows.

Graphs also help uncover how very small interactions and dynamics lead to global mutations. They tie together the micro and macro scales by representing exactly which things are interacting within global structures. These associations are used to forecast behavior and determine missing links. Figure 1-4 is a foodweb of grassland species interactions that used graph analysis to evaluate the hierarchical organization and species interactions and then predict missing relationships, as detailed in the paper by A. Clauset, C. Moore, and M. E. J. Newman, "Hierarchical Structure and the Prediction of Missing Links in Network" (*https://www.nature.com/articles/nature06830*).

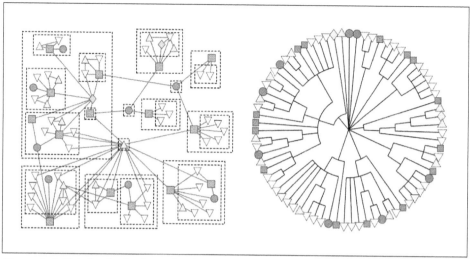

Figure 1-4. This foodweb of grassland species uses graphs to correlate small-scale interactions to larger structure formation.

Graph Processing, Databases, Queries, and Algorithms

Graph processing includes the methods by which graph workloads and tasks are carried out. Most graph queries consider specific parts of the graph (e.g., a starting node), and the work is usually focused in the surrounding subgraph. We term this type of work *graph local*, and it implies declaratively querying a graph's structure, as explained in the book *Graph Databases*, by Ian Robinson, Jim Webber, and Emil Eifrem (O'Reilly). This type of graph-local processing is often utilized for real-time transactions and pattern-based queries.

When speaking about graph algorithms, we are typically looking for global patterns and structures. The input to the algorithm is usually the whole graph, and the output can be an enriched graph or some aggregate value such as a score. We categorize such processing as *graph global*, and it implies processing a graph's structure using computational algorithms (often iteratively). This approach sheds light on the overall nature of a network through its connections. Organizations tend to use graph algorithms to model systems and predict behavior based on how things disseminate, important components, group identification, and the overall robustness of the system.

There may be some overlap in these definitions—sometimes we can use processing of an algorithm to answer a local query, or vice versa—but simplistically speaking whole-graph operations are processed by computational algorithms and subgraph operations are queried in databases.

Traditionally, transaction processing and analysis have been siloed. This was an unnatural split based on technology limitations. Our view is that graph analytics drives smarter transactions, which creates new data and opportunities for further analysis. More recently there's been a trend to integrate these silos for more real-time decision making.

OLTP and OLAP

Online transaction processing (OLTP) operations are typically short activities like booking a ticket, crediting an account, booking a sale, and so forth. OLTP implies voluminous low-latency query processing and high data integrity. Although OLTP may involve only a small number of records per transaction, systems process many transactions concurrently.

Online analytical processing (OLAP) facilitates more complex queries and analysis over historical data. These analyses may include multiple data sources, formats, and types. Detecting trends, conducting "what-if" scenarios, making predictions, and uncovering structural patterns are typical OLAP use cases. Compared to OLTP, OLAP systems process fewer but longer-running transactions over many records. OLAP systems are biased toward faster reading without the expectation of transactional updates found in OLTP, and batch-oriented operation is common.

Recently, however, the line between OLTP and OLAP has begun to blur. Modern data-intensive applications now combine real-time transactional operations with analytics. This merging of processing has been spurred by several advances in software, such as more scalable transaction management and incremental stream processing, and by lower-cost, large-memory hardware.

Bringing together analytics and transactions enables continual analysis as a natural part of regular operations. As data is gathered—from point-of-sale (POS) machines, manufacturing systems, or internet of things (IoT) devices—analytics now supports the ability to make real-time recommendations and decisions while processing. This trend was observed several years ago, and terms to describe this merging include *translytics* and *hybrid transactional and analytical processing* (HTAP). Figure 1-5 illustrates how read-only replicas can be used to bring together these different types of processing.

According to Gartner (*https://gtnr.it/2FAKnuX*):

> [HTAP] could potentially redefine the way some business processes are executed, as real-time advanced analytics (for example, planning, forecasting and what-if analysis) becomes an integral part of the process itself, rather than a separate activity performed after the fact. This would enable new forms of real-time business-driven decision-making process. Ultimately, HTAP will become a key enabling architecture for intelligent business operations.

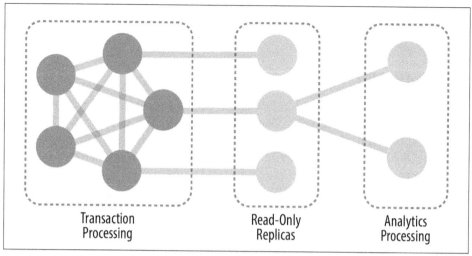

Figure 1-5. A hybrid platform supports the low latency query processing and high data integrity required for transactions while integrating complex analytics over large amounts of data.

As OLTP and OLAP become more integrated and begin to support functionality previously offered in only one silo, it's no longer necessary to use different data products or systems for these workloads—we can simplify our architecture by using the same platform for both. This means our analytical queries can take advantage of real-time data and we can streamline the iterative process of analysis.

Why Should We Care About Graph Algorithms?

Graph algorithms are used to help make sense of connected data. We see relationships within real-world systems from protein interactions to social networks, from communication systems to power grids, and from retail experiences to Mars mission planning. Understanding networks and the connections within them offers incredible potential for insight and innovation.

Graph algorithms are uniquely suited to understanding structures and revealing patterns in datasets that are highly connected. Nowhere is the connectivity and interactivity so apparent than in big data. The amount of information that has been brought together, commingled, and dynamically updated is impressive. This is where graph algorithms can help us to make sense of our volumes of data, with more sophisticated analytics that leverage relationships and enhance artificial intelligence contextual information.

As our data becomes more connected, it's increasingly important to understand its relationships and interdependencies. Scientists that study the growth of networks have noted that connectivity increases over time, but not uniformly. Preferential attachment is one theory on how the dynamics of growth impact structure. This idea, illustrated in Figure 1-6, describes the tendency of a node to link to other nodes that already have a lot of connections.

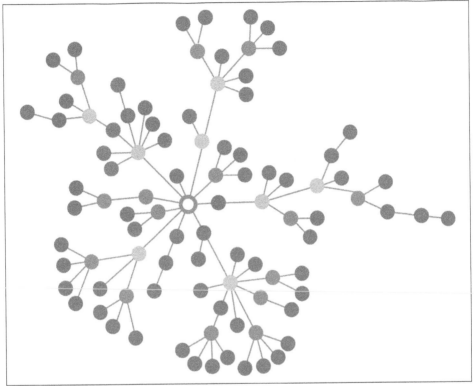

Figure 1-6. Preferential attachment is the phenomenon where the more connected a node is, the more likely it is to receive new links. This leads to uneven concentrations and hubs.

In his book, *Sync: How Order Emerges from Chaos in the Universe, Nature, and Daily Life* (Hachette), Steven Strogatz provides examples and explains different ways that real-life systems self-organize. Regardless of the underlying causes, many researchers believe that how networks grow is inseparable from their resulting shapes and hierarchies. Highly dense groups and lumpy data networks tend to develop, with complexity growing along with data size. We see this clustering of relationships in most real-world networks today, from the internet to social networks like the gaming community shown in Figure 1-7.

The network analysis shown in Figure 1-7 was created by Francesco D'Orazio of Pulsar to help predict the virality of content and inform distribution strategies. D'Orazio found (*https://bit.ly/2CCLlVI*) a correlation between the concentration of a community's distribution and the speed of diffusion of a piece of content.

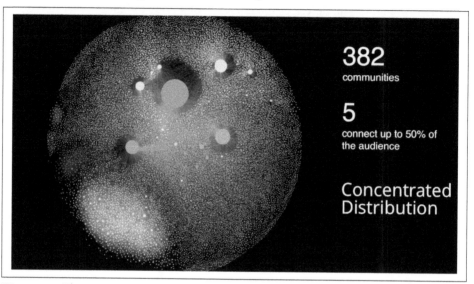

Figure 1-7. This gaming community analysis (https://bit.ly/2CCLlVI) shows a concentration of connections around just 5 of 382 communities.

This is significantly different than what an average distribution model would predict, where most nodes would have the same number of connections. For instance, if the World Wide Web had an average distribution of connections, all pages would have about the same number of links coming in and going out. Average distribution models assert that most nodes are equally connected, but many types of graphs and many real networks exhibit concentrations. The web, in common with graphs like travel and social networks, has a *power-law* distribution with a few nodes being highly connected and most nodes being modestly connected.

Power Law

A *power law* (also called a *scaling law*) describes the relationship between two quantities where one quantity varies as a power of another. For instance, the area of a cube is related to the length of its sides by a power of 3. A well-known example is the *Pareto distribution* or "80/20 rule," originally used to describe the situation where 20% of a population controlled 80% of the wealth. We see various power laws in the natural world and networks.

Trying to "average out" a network generally won't work well for investigating relationships or forecasting, because real-world networks have uneven distributions of nodes and relationships. We can readily see in Figure 1-8 how using an average of character-istics for data that is uneven would lead to incorrect results.

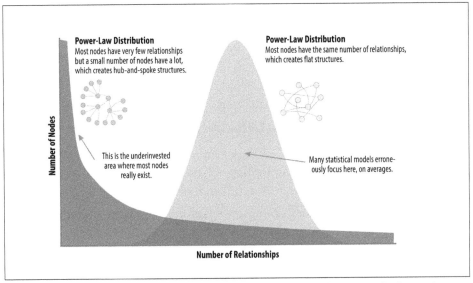

Figure 1-8. Real-world networks have uneven distributions of nodes and relationships represented in the extreme by a power-law distribution. An average distribution assumes most nodes have the same number of relationships and results in a random network.

Because highly connected data does not adhere to an average distribution, network scientists use graph analytics to search for and interpret structures and relationship distributions in real-world data.

> There is no network in nature that we know of that would be described by the random network model.
>
> —Albert-László Barabási, Director, Center for Complex Network Research, North-eastern University, and author of numerous network science books

The challenge for most users is that densely and unevenly connected data is trouble-some to analyze with traditional analytical tools. There might be a structure there, but it's hard to find. It's tempting to take an averages approach to messy data, but doing so will conceal patterns and ensure our results are not representing any real groups. For instance, if you average the demographic information of all your customers and offer an experience based solely on averages, you'll be guaranteed to miss most communi-ties: communities tend to cluster around related factors like age and occupation or marital status and location.

Furthermore, dynamic behavior, particularly around sudden events and bursts, can't be seen with a snapshot. To illustrate, if you imagine a social group with increasing relationships, you'd also expect more communications. This could lead to a tipping point of coordination and a subsequent coalition or, alternatively, subgroup formation and polarization in, for example, elections. Sophisticated methods are required to forecast a network's evolution over time, but we can infer behavior if we understand the structures and interactions within our data. Graph analytics is used to predict group resiliency because of the focus on relationships.

Graph Analytics Use Cases

At the most abstract level, graph analytics is applied to forecast behavior and prescribe action for dynamic groups. Doing this requires understanding the relationships and structure within the group. Graph algorithms accomplish this by examining the overall nature of networks through their connections. With this approach, you can understand the topology of connected systems and model their processes.

There are three general buckets of questions that indicate whether graph analytics and algorithms are warranted, as shown in Figure 1-9.

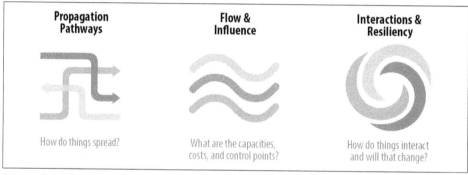

Propagation Pathways	Flow & Influence	Interactions & Resiliency
How do things spread?	What are the capacities, costs, and control points?	How do things interact and will that change?

Figure 1-9. The types of questions graph analytics answer

Here are a few types of challenges where graph algorithms are employed. Are your challenges similar?

- Investigate the route of a disease or a cascading transport failure.
- Uncover the most vulnerable, or damaging, components in a network attack.
- Identify the least costly or fastest way to route information or resources.
- Predict missing links in your data.
- Locate direct and indirect influence in a complex system.
- Discover unseen hierarchies and dependencies.
- Forecast whether groups will merge or break apart.
- Find bottlenecks or who has the power to deny/provide more resources.

- Reveal communities based on behavior for personalized recommendations.
- Reduce false positives in fraud and anomaly detection.
- Extract more predictive features for machine learning.

Conclusion

In this chapter, we've looked at how data today is extremely connected, and the implications of this. Robust scientific practices exist for analysis of group dynamics and relationships, yet those tools are not always commonplace in businesses. As we evaluate advanced analytics techniques, we should consider the nature of our data and whether we need to understand community attributes or predict complex behavior. If our data represents a network, we should avoid the temptation to reduce factors to an average. Instead, we should use tools that match our data and the insights we're seeking.

In the next chapter, we'll cover graph concepts and terminology.

CHAPTER 2

Graph Theory and Concepts

In this chapter, we set the framework and cover terminology for graph algorithms. The basics of graph theory are explained, with a focus on the concepts that are most relevant to a practitioner.

We'll describe how graphs are represented, and then explain the different types of graphs and their attributes. This will be important later, as our graph's characteristics will inform our algorithm choices and help us interpret results. We'll finish the chapter with an overview of the types of graph algorithms detailed in this book.

Terminology

The labeled property graph is one of the most popular ways of modeling graph data.

A *label* marks a node as part of a group. In Figure 2-1, we have two groups of nodes: Person and Car. (Although in classic graph theory a label applies to a single node, it's now commonly used to mean a node group.) Relationships are classified based on *relationship type*. Our example includes the relationship types of DRIVES, OWNS, LIVES_WITH, and MARRIED_TO.

Properties are synonymous with attributes and can contain a variety of data types, from numbers and strings to spatial and temporal data. In Figure 2-1 we assigned the properties as name-value pairs, where the name of the property comes first and then its value. For example, the Person node on the left has a property name: "Dan", and the MARRIED_TO relationship has a property of on: Jan 1, 2013.

A *subgraph* is a graph within a larger graph. Subgraphs are useful as a filters such as when we need a subset with particular characteristics for focused analysis.

A *path* is a group of nodes and their connecting relationships. An example of a simple path, based on Figure 2-1, could contain the nodes Dan, Ann, and Car and the DRIVES and OWNS relationships.

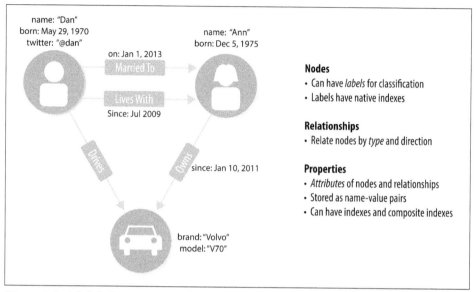

Figure 2-1. A labeled property graph model is a flexible and concise way of representing connected data.

Graphs vary in type, shape, and size as well the kind of attributes that can be used for analysis. Next, we'll describe the kinds of graphs most suited for graph algorithms. Keep in mind that these explanations apply to graphs as well as subgraphs.

Graph Types and Structures

In classic graph theory, the term *graph* is equated with a *simple* (or *strict*) graph where nodes only have one relationship between them, as shown on the left side of Figure 2-2. Most real-world graphs, however, have many relationships between nodes and even self-referencing relationships. Today, this term is commonly used for all three graph types in Figure 2-2, so we also use the term inclusively.

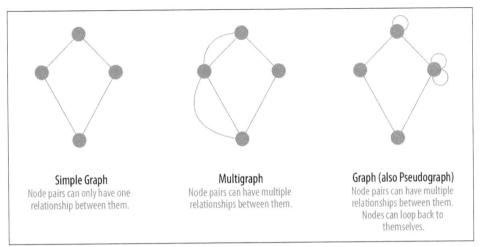

Figure 2-2. In this book, we use the term graph to include any of these classic types of graphs.

Random, Small-World, Scale-Free Structures

Graphs take on many shapes. Figure 2-3 shows three representative network types:

Random networks
> In a completely average distribution of connections, a *random network* is formed with no hierarchies. This type of shapeless graph is "flat" with no discernible patterns. All nodes have the same probability of being attached to any other node.

Small-world networks
> A *small-world network* is extremely common in social networks; it shows localized connections and some hub-and-spoke pattern. The "Six Degrees of Kevin Bacon" (*https://bit.ly/2FAbVk8*) game might be the best-known example of the small-world effect. Although you associate mostly with a small group of friends, you're never many hops away from anyone else—even if they are a famous actor or on the other side of the planet.

Scale-free networks
> A *scale-free network* is produced when there are power-law distributions and a hub-and-spoke architecture is preserved regardless of scale, such as in the World Wide Web.

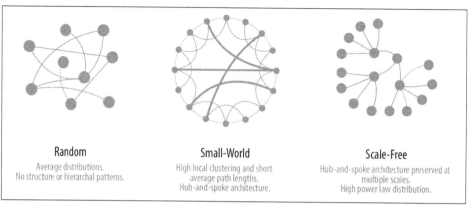

Random
Average distributions.
No structure or hierarchal patterns.

Small-World
High local clustering and short
average path lengths.
Hub-and-spoke architecture.

Scale-Free
Hub-and-spoke architecture preserved at
multiple scales.
High power law distribution.

Figure 2-3. Three network structures with distinctive graphs and behaviors

These network types produce graphs with distinctive structures, distributions, and behaviors. As we work with graph algorithms, we'll come to recognize similar patterns in our results.

Flavors of Graphs

To get the most out of graph algorithms, it's important to familiarize ourselves with the most characteristic graphs we'll encounter. Table 2-1 summarizes common graph attributes. In the following sections we look at the different flavors in more detail.

Table 2-1. Common attributes of graphs

Graph attribute	Key factor	Algorithm consideration
Connected versus disconnected	Whether there is a path between any two nodes in the graph, irrespective of distance	Islands of nodes can cause unexpected behavior, such as getting stuck in or failing to process disconnected components.
Weighted versus unweighted	Whether there are (domain-specific) values on relationships or nodes	Many algorithms expect weights, and we'll see significant differences in performance and results when they're ignored.
Directed versus undirected	Whether or not relationships explicitly define a start and end node	This adds rich context to infer additional meaning. In some algorithms you can explicitly set the use of one, both, or no direction.
Cyclic versus acyclic	Whether paths start and end at the same node	Cyclic graphs are common but algorithms must be careful (typically by storing traversal state) or cycles may prevent termination. Acyclic graphs (or spanning trees) are the basis for many graph algorithms.
Sparse versus dense	Relationship to node ratio	Extremely dense or extremely sparsely connected graphs can cause divergent results. Data modeling may help, assuming the domain is not inherently dense or sparse.

Graph attribute	Key factor	Algorithm consideration
Monopartite, bipartite, and k-partite	Whether nodes connect to only one other node type (e.g., users like movies) or many other node types (e.g., users like users who like movies)	Helpful for creating relationships to analyze and projecting more useful graphs.

Connected Versus Disconnected Graphs

A graph is connected if there is a path between all nodes. If we have islands in our graph, it's disconnected. If the nodes in those islands are connected, they are called *components* (or sometimes *clusters*), as shown in Figure 2-4.

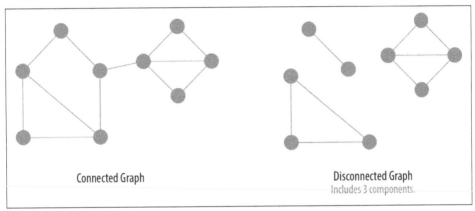

Figure 2-4. *If we have islands in our graph, it's a disconnected graph.*

Some algorithms struggle with disconnected graphs and can produce misleading results. If we have unexpected results, checking the structure of our graph is a good first step.

Unweighted Graphs Versus Weighted Graphs

Unweighted graphs have no weight values assigned to their nodes or relationships. For weighted graphs, these values can represent a variety of measures such as cost, time, distance, capacity, or even a domain-specific prioritization. Figure 2-5 visualizes the difference.

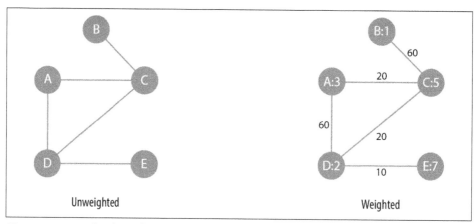

Figure 2-5. Weighted graphs can hold values on relationships or nodes.

Basic graph algorithms can use weights for processing as a representation for the strength or value of relationships. Many algorithms compute metrics which can then be used as weights for follow-up processing. Some algorithms update weight values as they proceed to find cumulative totals, lowest values, or optimums.

A classic use for weighted graphs is in pathfinding algorithms. Such algorithms underpin the mapping applications on our phones and compute the shortest/cheapest/fastest transport routes between locations. For example, Figure 2-6 uses two different methods of computing the shortest route.

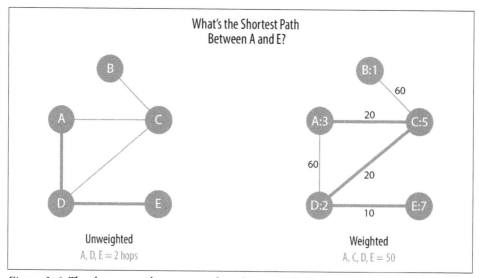

Figure 2-6. The shortest paths can vary for otherwise identical unweighted and weighted graphs.

Without weights, our shortest route is calculated in terms of the number of relationships (commonly called *hops*). A and E have a two-hop shortest path, which indicates only one node (D) between them. However, the shortest weighted path from A to E takes us from A to C to D to E. If weights represent a physical distance in kilometers, the total distance would be 50 km. In this case, the shortest path in terms of the number of hops would equate to a longer physical route of 70 km.

Undirected Graphs Versus Directed Graphs

In an undirected graph, relationships are considered bidirectional (for example, friendships). In a directed graph, relationships have a specific direction. Relationships pointing to a node are referred to as *in-links* and, unsurprisingly, *out-links* are those originating from a node.

Direction adds another dimension of information. Relationships of the same type but in opposing directions carry different semantic meaning, expressing a dependency or indicating a flow. This may then be used as an indicator of credibility or group strength. Personal preferences and social relations are expressed very well with direction.

For example, if we assumed in Figure 2-7 that the directed graph was a network of students and the relationships were "likes," then we'd calculate that A and C are more popular.

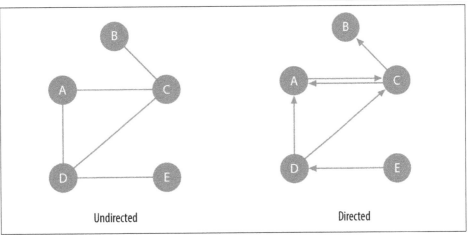

Undirected   Directed

Figure 2-7. Many algorithms allow us to compute on the basis of only inbound or outbound connections, both directions, or without direction.

Road networks illustrate why we might want to use both types of graphs. For example, highways between cities are often traveled in both directions. However, within cities, some roads are one-way streets. (The same is true for some information flows!)

We get different results running algorithms in an undirected fashion compared to directed. In an undirected graph, for example for highways or friendships, we would assume all relationships always go both ways.

If we reimagine Figure 2-7 as a directed road network, you can drive to A from C and D but you can only leave through C. Furthermore if there were no relationships from A to C, that would indicate a dead end. Perhaps that's less likely for a one-way road network, but not for a process or a web page.

Acyclic Graphs Versus Cyclic Graphs

In graph theory, *cycles* are paths through relationships and nodes that start and end at the same node. An *acyclic graph* has no such cycles. As shown in Figure 2-8, both directed and undirected graphs can have cycles, but when directed, paths follow the relationship direction. A *directed acyclic graph* (DAG), shown in Graph 1, will by definition always have dead ends (also called *leaf nodes*).

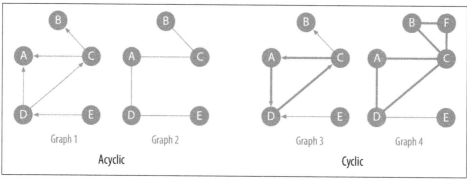

Figure 2-8. In acyclic graphs, it's impossible to start and end on the same node without retracing our steps.

Graphs 1 and 2 have no cycles, as there's no way to start and end at the same node without repeating a relationship. You might remember from Chapter 1 that not repeating relationships was the Königsberg bridges problem that started graph theory! Graph 3 in Figure 2-8 shows a simple cycle following A-D-C-A with no repeated nodes. In Graph 4, the undirected cyclic graph has been made more interesting by adding a node and relationship. There's now a closed cycle with a repeated node (C), following B-F-C-D-A-C-B. There are actually multiple cycles in Graph 4.

Cycles are common, and we sometimes need to convert cyclic graphs to acyclic graphs (by cutting relationships) to eliminate processing problems. Directed acyclic graphs naturally arise in scheduling, genealogy, and version histories.

Trees

In classic graph theory, an acyclic graph that is undirected is called a *tree*. In computer science, trees can also be directed. A more inclusive definition would be a graph where any two nodes are connected by only one path. Trees are significant for understanding graph structures and many algorithms. They play a key role in designing networks, data structures, and search optimizations to improve categorization or organizational hierarchies.

Much has been written about trees and their variations. Figure 2-9 illustrates the common trees that we're likely to encounter.

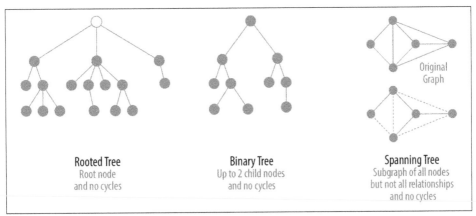

Figure 2-9. Of these prototypical tree graphs, spanning trees are most often used for graph algorithms.

Of these variations, spanning trees are the most relevant for this book. A *spanning tree* is an acyclic subgraph that includes all the nodes of a larger graph but not all the relationships. A minimum spanning tree connects all the nodes of a graph with either the least number of hops or least weighted paths.

Sparse Graphs Versus Dense Graphs

The sparsity of a graph is based on the number of relationships it has compared to the maximum possible number of relationships, which would occur if there was a relationship between every pair of nodes. A graph where every node has a relationship with every other node is called a *complete graph*, or a *clique* for components. For instance, if all my friends knew each other, that would be a clique.

The *maximum density* of a graph is the number of relationships possible in a complete graph. It's calculated with the formula $MaxD = \frac{N(N-1)}{2}$ where N is the number of nodes. To measure *actual density* we use the formula $D = \frac{2(R)}{N(N-1)}$ where R is the number of relationships. In Figure 2-10, we can see three measures of actual density for undirected graphs.

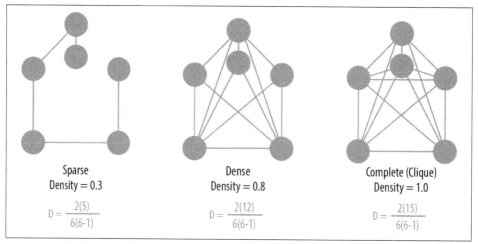

Figure 2-10. Checking the density of a graph can help you evaluate unexpected results.

Although there is no strict dividing line, any graph with an actual density that approaches the maximum density is considered dense. Most graphs based on real networks tend toward sparseness, with an approximately linear correlation of total nodes to total relationships. This is especially the case where physical elements come into play, such as the practical limitations to how many wires, pipes, roads, or friendships you can join at one point.

Some algorithms will return nonsensical results when executed on extremely sparse or dense graphs. If a graph is too sparse there may not be enough relationships for algorithms to compute useful results. Alternatively, very densely connected nodes don't add much additional information since they are so highly connected. High densities can also skew some results or add computational complexity. In these situations, filtering out the relevant subgraph is a practical approach.

Monopartite, Bipartite, and k-Partite Graphs

Most networks contain data with multiple node and relationship types. Graph algorithms, however, frequently consider only one node type and one relationship type. Graphs with one node type and relationship type are sometimes referred to as *monopartite*.

A *bipartite* graph is a graph whose nodes can be divided into two sets, such that relationships only connect a node from one set to a node from a different set. Figure 2-11 shows an example of such a graph. It has two sets of nodes: a viewer set and a TV show set. There are only relationships between the two sets and no intraset connections. In other words in Graph 1, TV shows are only related to viewers, not other TV shows, and viewers are likewise not directly linked to other viewers.

Starting from our bipartite graph of viewers and TV shows, we created two monopartite projections: Graph 2 of viewer connections based on shows in common, and Graph 3 of TV shows based on viewers in common. We can also filter based on relationship type, such as watched, rated, or reviewed.

Projecting monopartite graphs with inferred connections is an important part of graph analysis. These types of projections help uncover indirect relationships and qualities. For example, in Graph 2 in Figure 2-11, Bev and Ann have watched only one TV show in common whereas Bev and Evan have two shows in common. In Graph 3 we've weighted the relationships between the TV shows by the aggregated views by viewers in common. This, or other metrics such as similarity, can be used to infer meaning between activities like watching *Battlestar Galactica* and *Firefly*. That can inform our recommendation for someone similar to Evan who, in Figure 2-11, just finished watching the last episode of *Firefly*.

k-partite graphs reference the number of node types our data has (k). For example, if we have three node types, we'd have a tripartite graph. This just extends bipartite and monopartite concepts to account for more node types. Many real-world graphs, especially knowledge graphs, have a large value for k, as they combine many different concepts and types of information. An example of using a larger number of node types is creating new recipes by mapping a recipe set to an ingredient set to a chemical compound, and then deducing new mixes that connect popular preferences. We could also reduce the number of nodes types by generalization, such as treating many forms of a node, like spinach or collards, as just "leafy greens."

Now that we've reviewed the types of graphs we're most likely to work with, let's learn about the types of graph algorithms we'll execute on those graphs.

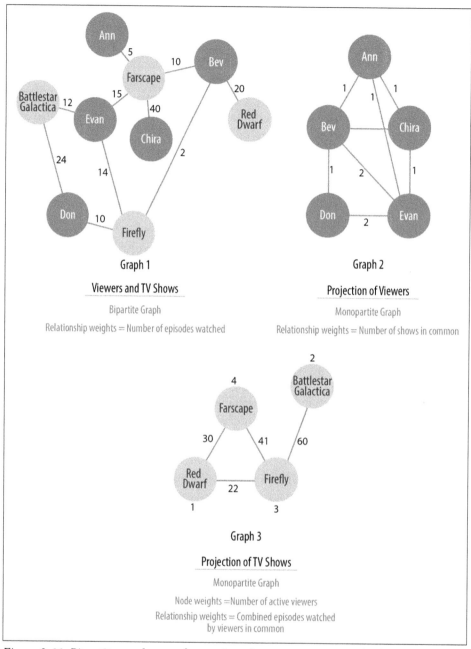

Figure 2-11. Bipartite graphs are often projected to monopartite graphs for more specific analysis.

Types of Graph Algorithms

Let's look into the three areas of analysis that are at the heart of graph algorithms. These categories correspond to the chapters on algorithms for pathfinding and search, centrality computation, and community detection.

Pathfinding

Paths are fundamental to graph analytics and algorithms, so this is where we'll start our chapters with specific algorithm examples. Finding shortest paths is probably the most frequent task performed with graph algorithms and is a precursor for several different types of analysis. The shortest path is the traversal route with the fewest hops or lowest weight. If the graph is directed, then it's the shortest path between two nodes as allowed by the relationship directions.

Path Types

The *average shortest path* is used to consider the overall efficiency and resiliency of networks, such as understanding the average distance between subway stations. Sometimes we may also want to understand the longest optimized route for situations such as determining which subway stations are the farthest apart or have the most number of stops between them even when the best route is chosen. In this case, we'd use the *diameter* of a graph to find the longest shortest path between all node pairs.

Centrality

Centrality is all about understanding which nodes are more important in a network. But what do we mean by importance? There are different types of centrality algorithms created to measure different things, such as the ability to quickly spread information versus bridging distinct groups. In this book, we'll focus on how nodes and relationships are structured.

Community Detection

Connectedness is a core concept of graph theory that enables a sophisticated network analysis such as finding communities. Most real-world networks exhibit substructures (often quasi-fractal) of more or less independent subgraphs.

Connectivity is used to find communities and quantify the quality of groupings. Evaluating different types of communities within a graph can uncover structures, like hubs and hierarchies, and tendencies of groups to attract or repel others. These techniques are used to study emergent phenomena such as those that lead to echo chambers and filter bubble effects.

Summary

Graphs are intuitive. They align with how we think about and draw systems. The primary tenets of working with graphs can be quickly assimilated once we've unraveled some of the terminology and layers. In this chapter we've explained the ideas and expressions used later in this book and described flavors of graphs you'll come across.

Graph Theory References

If you're excited to learn more about graph theory itself, there are a few introductory texts we recommend:

- *Introduction to Graph Theory*, by Richard J. Trudeau (Dover), is a very well written, gentle introduction.
- *Introduction to Graph Theory*, Fifth Ed., by Robin J. Wilson (Pearson), is a solid introduction with good illustrations.
- *Graph Theory and Its Applications*, Third Ed., by Jonathan L. Gross, Jay Yellen, and Mark Anderson (Chapman and Hall), assumes more mathematics background and provides more detail and exercises.

Next, we'll look at graph processing and types of analysis before diving into how to use graph algorithms in Apache Spark and Neo4j.

CHAPTER 3

Graph Platforms and Processing

In this chapter, we'll quickly cover different methods for graph processing and the most common platform approaches. We'll look more closely at the two platforms used in this book, Apache Spark and Neo4j, and when they may be appropriate for different requirements. Platform installation guidelines are included to prepare you for the next several chapters.

Graph Platform and Processing Considerations

Graph analytical processing has unique qualities such as computation that is structure-driven, globally focused, and difficult to parse. In this section we'll look at the general considerations for graph platforms and processing.

Platform Considerations

There's debate as to whether it's better to scale up or scale out graph processing. Should you use powerful multicore, large-memory machines and focus on efficient data structures and multithreaded algorithms? Or are investments in distributed processing frameworks and related algorithms worthwhile?

A useful evaluation approach is the *Configuration that Outperforms a Single Thread* (COST), as described in the research paper "Scalability! But at What COST?" (*https://bit.ly/2Ypjhyv*) by F. McSherry, M. Isard, and D. Murray. COST provides us with a way to compare a system's scalability with the overhead the system introduces. The core concept is that a well-configured system using an optimized algorithm and data structure can outperform current general-purpose scale-out solutions. It's a method for measuring performance gains without rewarding systems that mask inefficiencies through parallelization. Separating the ideas of scalability and efficient use of resources will help us build a platform configured explicitly for our needs.

Some approaches to graph platforms include highly integrated solutions that optimize algorithms, processing, and memory retrieval to work in tighter coordination.

Processing Considerations

There are different approaches for expressing data processing; for example, stream or batch processing or the map-reduce paradigm for records-based data. However, for graph data, there also exist approaches which incorporate the data dependencies inherent in graph structures into their processing:

Node-centric

>This approach uses nodes as processing units, having them accumulate and compute state and communicate state changes via messages to their neighbors. This model uses the provided transformation functions for more straightforward implementations of each algorithm.

Relationship-centric

>This approach has similarities with the node-centric model but may perform better for subgraph and sequential analysis.

Graph-centric

>These models process nodes within a subgraph independently of other subgraphs while (minimal) communication to other subgraphs happens via messaging.

Traversal-centric

>These models use the accumulation of data by the traverser while navigating the graph as their means of computation.

Algorithm-centric

>These approaches use various methods to optimize implementations per algorithm. This is a hybrid of the previous models.

 Pregel (https://bit.ly/2Twj9sY) is a node-centric, fault-tolerant parallel processing framework created by Google for performant analysis of large graphs. Pregel is based on the *bulk synchronous parallel* (BSP) model. BSP simplifies parallel programming by having distinct computation and communication phases.

Pregel adds a node-centric abstraction atop BSP whereby algorithms compute values from incoming messages from each node's neighbors. These computations are executed once per iteration and can update node values and send messages to other nodes. The nodes can also combine messages for transmission during the communication phase, which helpfully reduces the amount of network chatter. The algorithm completes when either no new messages are sent or a set limit has been reached.

Most of these graph-specific approaches require the presence of the entire graph for efficient cross-topological operations. This is because separating and distributing the graph data leads to extensive data transfers and reshuffling between worker instances. This can be difficult for the many algorithms that need to iteratively process the global graph structure.

Representative Platforms

To address the requirements of graph processing, several platforms have emerged. Traditionally there was a separation between graph compute engines and graph databases, which required users to move their data depending on their process needs:

Graph compute engines
These are read-only, nontransactional engines that focus on efficient execution of iterative graph analytics and queries of the whole graph. Graph compute engines support different definition and processing paradigms for graph algorithms, like node-centric (e.g., Pregel, Gather-Apply-Scatter) or MapReduce-based approaches (e.g., PACT). Examples of such engines are Giraph, GraphLab, Graph-Engine, and Apache Spark.

Graph databases
From a transactional background, these focus on fast writes and reads using smaller queries that generally touch a small fraction of a graph. Their strengths are in operational robustness and high concurrent scalability for many users.

Selecting Our Platform

Choosing a production platform involves many considersations, such as the type of analysis to be run, performance needs, the existing environment, and team preferences. We use Apache Spark and Neo4j to showcase graph algorithms in this book because they both offer unique advantages.

Spark is an example of a scale-out and node-centric graph compute engine. Its popular computing framework and libraries support a variety of data science workflows. Spark may be the right platform when our:

- Algorithms are fundamentally parallelizable or partitionable.
- Algorithm workflows need "multilingual" operations in multiple tools and languages.
- Analysis can be run offline in batch mode.
- Graph analysis is on data not transformed into a graph format.
- Team needs and has the expertise to code and implement their own algorithms.
- Team uses graph algorithms infrequently.

- Team prefers to keep all data and analysis within the Hadoop ecosystem.

The Neo4j Graph Platform is an example of a tightly integrated graph database and algorithm-centric processing, optimized for graphs. It is popular for building graph-based applications and includes a graph algorithms library tuned for its native graph database. Neo4j may be the right platform when our:

- Algorithms are more iterative and require good memory locality.
- Algorithms and results are performance sensitive.
- Graph analysis is on complex graph data and/or requires deep path traversal.
- Analysis/results are integrated with transactional workloads.
- Results are used to enrich an existing graph.
- Team needs to integrate with graph-based visualization tools.
- Team prefers prepackaged and supported algorithms.

Finally, some organizations use both Neo4j and Spark for graph processing: Spark for the high-level filtering and preprocessing of massive datasets and data integration, and Neo4j for more specific processing and integration with graph-based applications.

Apache Spark

Apache Spark (henceforth just Spark) is an analytics engine for large-scale data processing. It uses a table abstraction called a DataFrame to represent and process data in rows of named and typed columns. The platform integrates diverse data sources and supports languages such as Scala, Python, and R. Spark supports various analytics libraries, as shown in Figure 3-1. Its memory-based system operates by using efficiently distributed compute graphs.

GraphFrames is a graph processing library for Spark that succeeded GraphX in 2016, although it is separate from the core Apache Spark. GraphFrames is based on GraphX, but uses DataFrames as its underlying data structure. GraphFrames has support for the Java, Scala, and Python programming languages. In this book, our examples will be based on the Python API (PySpark) because of its current popularity with Spark data scientists.

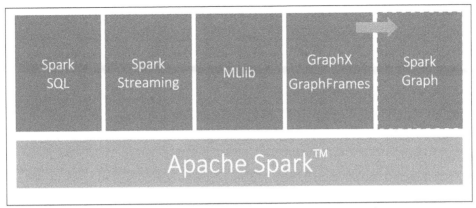

Figure 3-1. Spark is an open-source distributed and general-purpose clustercomputing framework. It includes several modules for various workloads.

Nodes and relationships are represented as DataFrames with a unique ID for each node and a source and destination node for each relationship. We can see an example of a nodes DataFrame in Table 3-1 and a relationships DataFrame in Table 3-2. A GraphFrame based on these DataFrames would have two nodes, JFK and SEA, and one relationship, from JFK to SEA.

Table 3-1. Nodes DataFrame

id	city	state
JFK	New York	NY
SEA	Seattle	WA

Table 3-2. Relationships DataFrame

src	dst	delay	tripId
JFK	SEA	45	1058923

The nodes DataFrame must have an id column—the value in this column is used to uniquely identify each node. The relationships DataFrame must have src and dst columns—the values in these columns describe which nodes are connected and should refer to entries that appear in the id column of the nodes DataFrame.

The nodes and relationships DataFrames can be loaded using any of the DataFrame data sources (*http://bit.ly/2CN7LDV*), including Parquet, JSON, and CSV. Queries are described using a combination of the PySpark API and Spark SQL.

GraphFrames also provides users with an extension point (*http://bit.ly/2Wo6Hxg*) to implement algorithms that aren't available out of the box.

Installing Spark

You can download Spark from the Apache Spark website (*http://bit.ly/1qnQ5zb*). Once you've downloaded it, you need to install the following libraries to execute Spark jobs from Python:

```
pip install pyspark graphframes
```

You can then launch the *pyspark* REPL by executing the following command:

```
export SPARK_VERSION="spark-2.4.0-bin-hadoop2.7"
./${SPARK_VERSION}/bin/pyspark \
    --driver-memory 2g \
    --executor-memory 6g \
    --packages graphframes:graphframes:0.7.0-spark2.4-s_2.11
```

At the time of writing the latest released version of Spark is *spark-2.4.0-bin-hadoop2.7*, but that may have changed by the time you read this. If so, be sure to change the SPARK_VERSION environment variable appropriately.

 Although Spark jobs should be executed on a cluster of machines, for demonstration purposes we're only going to execute the jobs on a single machine. You can learn more about running Spark in production environments in *Spark: The Definitive Guide*, by Bill Chambers and Matei Zaharia (O'Reilly).

You're now ready to learn how to run graph algorithms on Spark.

Neo4j Graph Platform

The Neo4j Graph Platform supports transactional processing and analytical processing of graph data. It includes graph storage and compute with data management and analytics tooling. The set of integrated tools sits on top of a common protocol, API, and query language (Cypher) to provide effective access for different uses, as shown in Figure 3-2.

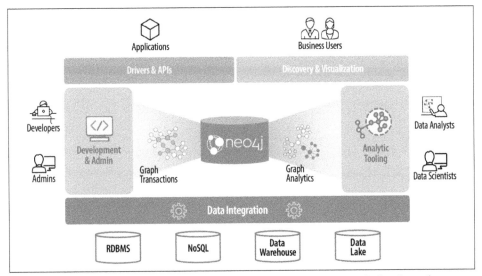

Figure 3-2. The Neo4j Graph Platform is built around a native graph database that supports transactional applications and graph analytics.

In this book, we'll be using the Neo4j Graph Data Science library (*https://neo4j.com/docs/graph-data-science/current/*). The library is installed as a plug-in alongside the database and provides a set of user-defined procedures (*https://bit.ly/2OmidGK*) that can be executed via the Cypher query language.

The Graph Data Science library includes parallel versions of algorithms supporting graph analytics and machine learning workflows. The algorithms are executed on top of a task-based parallel computation framework and are optimized for the Neo4j platform. For different graph sizes there are internal implementations that scale up to tens of billions of nodes and relationships.

Results can be streamed to the client as a tuples stream and tabular results can be used as a driving table for further processing. Results can also be optionally written back to the database efficiently as node properties or relationship types.

> In this book, we'll also be using the Neo4j Awesome Procedures on Cypher (APOC) library (*https://bit.ly/2JDfSbS*). APOC holds more than 450 procedures and functions that help with common tasks such as data integration, data conversion, and model refactoring.

Installing Neo4j

Neo4j Desktop is a convenient way for developers to work with local Neo4j databases. It can be downloaded from the Neo4j website (*https://neo4j.com/download/*). The graph algorithms and APOC libraries can be installed as plug-ins once you've installed and launched the Neo4j Desktop. In the lefthand menu, create a project and select it. Then click Manage on the database where you want to install the plug-ins. On the Plugins tab, you'll see options for several plug-ins. Click the Install button for graph algorithms and APOC. See Figures 3-3 and 3-4.

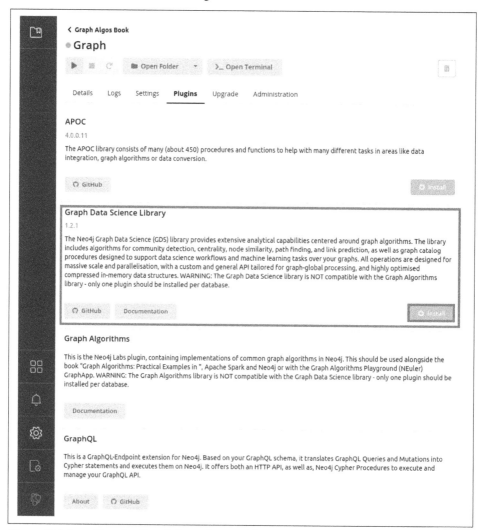

Figure 3-3. Installing the Graph Data Science library

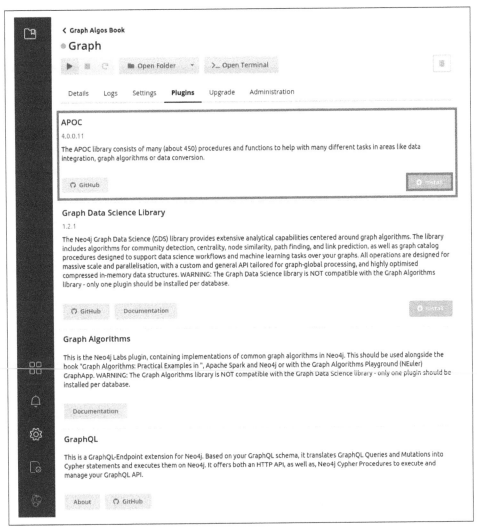

Figure 3-4. Installing the APOC library

Jennifer Reif explains the installation process in more detail in her blog post "Explore New Worlds—Adding Plugins to Neo4j" (*https://bit.ly/2TU0Lj3*). You're now ready to learn how to run graph algorithms in Neo4j.

The Neo4j Graph Data Science library contains many graph algorithms. The algorithms are divided into three maturity tiers indicated by a prefix of *gds*, *gds.beta* or *gds.alpha*. Over time the algorithms will continue to graduate to more mature tiers. If you find an algorithm can't be called from the lower tier indicated in the book, try the next level of maturity.

- The *production-quality tier* indicates that the algorithm has been thoroughly tested for high stability and scalability. Algorithms in this tier are prefixed with `gds.<algorithm>`.

- The *beta tier* indicates that the algorithm is a candidate for the production-quality tier but is still going through testing. Algorithms in this tier are prefixed with `gds.beta.<algorithm>`.

- The *alpha tier* indicates that the algorithm is experimental and might be changed or removed. Algorithms in this tier are prefixed with `gds.alpha.<algorithm>`.

Summary

In the previous chapters we've described why graph analytics is important to studying real-world networks and looked at fundamental graph concepts, analysis, and processing. This puts us on solid footing for understanding how to apply graph algorithms. In the next chapters, we'll discover how to run graph algorithms with examples in Spark and Neo4j.

CHAPTER 4

Pathfinding and Graph Search Algorithms

Graph search algorithms explore a graph either for general discovery or explicit search. These algorithms carve paths through the graph, but there is no expectation that those paths are computationally optimal. We will cover Breadth First Search and Depth First Search because they are fundamental for traversing a graph and are often a required first step for many other types of analysis.

Pathfinding algorithms build on top of graph search algorithms and explore routes between nodes, starting at one node and traversing through relationships until the destination has been reached. These algorithms are used to identify optimal routes through a graph for uses such as logistics planning, least cost call or IP routing, and gaming simulation.

Specifically, the pathfinding algorithms we'll cover are:

Shortest Path, with two useful variations (A* and Yen's)
    Finding the shortest path or paths between two chosen nodes

All Pairs Shortest Path and Single Source Shortest Path
    For finding the shortest paths between all pairs or from a chosen node to all others

Minimum Spanning Tree
    For finding a connected tree structure with the smallest cost for visiting all nodes from a chosen node

Random Walk
    Because it's a useful preprocessing/sampling step for machine learning workflows and other graph algorithms

In this chapter we'll explain how these algorithms work and show examples in Spark and Neo4j. In cases where an algorithm is only available in one platform, we'll provide just that one example or illustrate how you can customize our implementation.

Figure 4-1 shows the key differences between these types of algorithms, and Table 4-1 is a quick reference to what each algorithm computes with an example use.

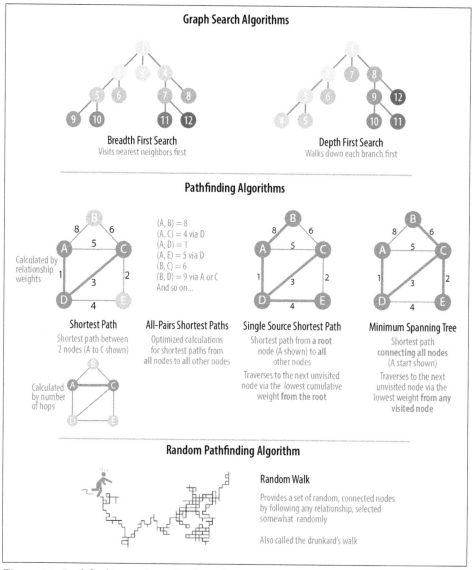

Figure 4-1. Pathfinding and search algorithms

Table 4-1. Overview of pathfinding and graph search algorithms

Algorithm type	What it does	Example use	Spark example	Neo4j example
Breadth First Search	Traverses a tree structure by fanning out to explore the nearest neighbors and then their sublevel neighbors	Locating neighbor nodes in GPS systems to identify nearby places of interest	Yes	No
Depth First Search	Traverses a tree structure by exploring as far as possible down each branch before backtracking	Discovering an optimal solution path in gaming simulations with hierarchical choices	No	No
Shortest Path Variations: A*, Yen's	Calculates the shortest path between a pair of nodes	Finding driving directions between two locations	Yes	Yes
All Pairs Shortest Path	Calculates the shortest path between *all pairs of nodes* in the graph	Evaluating alternate routes around a traffic jam	Yes	Yes
Single Source Shortest Path	Calculates the shorest path between a *single root* node and *all* other nodes	Least cost routing of phone calls	Yes	Yes
Minimum Spanning Tree	Calculates the path in a connected tree structure with the smallest cost for visiting all nodes	Optimizing connected routing, such as laying cable or garbage collection	No	Yes
Random Walk	Returns a list of nodes along a path of specified size by randomly choosing relationships to traverse.	Augmenting training for machine learning or data for graph algorithms.	No	Yes

First we'll take a look at the dataset for our examples and walk through how to import the data into Apache Spark and Neo4j. For each algorithm, we'll start with a short description of the algorithm and any pertinent information on how it operates. Most sections also include guidance on when to use related algorithms. Finally, we provide working sample code using the sample dataset at the end of each algorithm section.

Let's get started!

Example Data: The Transport Graph

All connected data contains paths between nodes, which is why search and pathfinding are the starting points for graph analytics. Transportation datasets illustrate these relationships in an intuitive and accessible way. The examples in this chapter run against a graph containing a subset of the European road network (*http://www.elbruz.org/e-roads/*). You can download the nodes and relationships files from the book's GitHub repository (*https://bit.ly/2FPgGVV*).

Table 4-2. transport-nodes.csv

id	latitude	longitude	population
Amsterdam	52.379189	4.899431	821752
Utrecht	52.092876	5.104480	334176
Den Haag	52.078663	4.288788	514861
Immingham	53.61239	-0.22219	9642
Doncaster	53.52285	-1.13116	302400
Hoek van Holland	51.9775	4.13333	9382
Felixstowe	51.96375	1.3511	23689
Ipswich	52.05917	1.15545	133384
Colchester	51.88921	0.90421	104390
London	51.509865	-0.118092	8787892
Rotterdam	51.9225	4.47917	623652
Gouda	52.01667	4.70833	70939

Table 4-3. transport-relationships.csv

src	dst	relationship	cost
Amsterdam	Utrecht	EROAD	46
Amsterdam	Den Haag	EROAD	59
Den Haag	Rotterdam	EROAD	26
Amsterdam	Immingham	EROAD	369
Immingham	Doncaster	EROAD	74
Doncaster	London	EROAD	277
Hoek van Holland	Den Haag	EROAD	27
Felixstowe	Hoek van Holland	EROAD	207
Ipswich	Felixstowe	EROAD	22
Colchester	Ipswich	EROAD	32
London	Colchester	EROAD	106
Gouda	Rotterdam	EROAD	25
Gouda	Utrecht	EROAD	35
Den Haag	Gouda	EROAD	32
Hoek van Holland	Rotterdam	EROAD	33

Figure 4-2 shows the target graph that we want to construct.

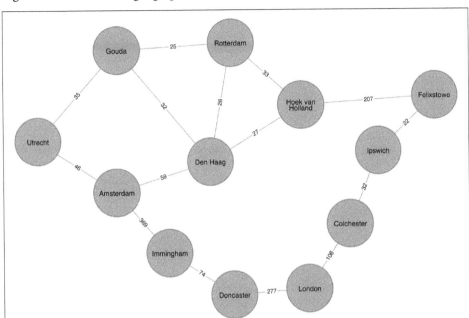

Figure 4-2. The transport graph

For simplicity we consider the graph in Figure 4-2 to be undirected because most roads between cities are bidirectional. We'd get slightly different results if we evaluated the graph as directed because of the small number of one-way streets, but the overall approach remains similar. However, both Spark and Neo4j operate on directed graphs. In cases like this where we want to work with undirected graphs (e.g., bidirectional roads), there is an easy way to accomplish that:

- For Spark, we'll create two relationships for each row in *transport-relationships.csv*—one going from dst to src and one from src to dst.

- For Neo4j, we'll create a single relationship and then ignore the relationship direction when we run the algorithms.

Having understood those little modeling workarounds, we can now get on with loading graphs into Spark and Neo4j from the example CSV files.

Importing the Data into Apache Spark

Starting with Spark, we'll first import the packages we need from Spark and the GraphFrames package:

```
from pyspark.sql.types import *
from graphframes import *
```

The following function creates a GraphFrame from the example CSV files:

```
def create_transport_graph():
    node_fields = [
        StructField("id", StringType(), True),
        StructField("latitude", FloatType(), True),
        StructField("longitude", FloatType(), True),
        StructField("population", IntegerType(), True)
    ]
    nodes = spark.read.csv("data/transport-nodes.csv", header=True,
                            schema=StructType(node_fields))

    rels = spark.read.csv("data/transport-relationships.csv", header=True)
    reversed_rels = (rels.withColumn("newSrc", rels.dst)
                    .withColumn("newDst", rels.src)
                    .drop("dst", "src")
                    .withColumnRenamed("newSrc", "src")
                    .withColumnRenamed("newDst", "dst")
                    .select("src", "dst", "relationship", "cost"))

    relationships = rels.union(reversed_rels)

    return GraphFrame(nodes, relationships)
```

Loading the nodes is easy, but for the relationships we need to do a little preprocessing so that we can create each relationship twice.

Now let's call that function:

```
g = create_transport_graph()
```

Importing the Data into Neo4j

Now for Neo4j. We'll start by creating a database that we'll use for the examples in this chapter:

```
:use system; ❶
CREATE DATABASE chapter4; ❷
:use chapter4; ❸
```

❶ Switch to the system database.

❷ Create a new database with the name chapter4. This operation is asynchronous so we may have to wait a couple of seconds before switching to the database.

❸ Switch to the chapter4 database.

Now let's load the nodes:

```
WITH 'https://github.com/neo4j-graph-analytics/book/raw/master/data/' AS base
WITH base + 'transport-nodes.csv' AS uri
LOAD CSV WITH HEADERS FROM uri  AS row
MERGE (place:Place {id:row.id})
SET place.latitude = toFloat(row.latitude),
    place.longitude = toFloat(row.longitude),
    place.population = toInteger(row.population);
```

And now the relationships:

```
WITH 'https://github.com/neo4j-graph-analytics/book/raw/master/data/' AS base
WITH base + 'transport-relationships.csv' AS uri
LOAD CSV WITH HEADERS FROM uri AS row
MATCH (origin:Place {id: row.src})
MATCH (destination:Place {id: row.dst})
MERGE (origin)-[:EROAD {distance: toInteger(row.cost)}]->(destination);
```

Although we're storing directed relationships, we'll ignore the direction when we execute algorithms later in the chapter.

Breadth First Search

Breadth First Search (BFS) is one of the fundamental graph traversal algorithms. It starts from a chosen node and explores all of its neighbors at one hop away before visiting all the neighbors at two hops away, and so on.

The algorithm was first published in 1959 by Edward F. Moore, who used it to find the shortest path out of a maze. It was then developed into a wire routing algorithm by C. Y. Lee in 1961, as described in "An Algorithm for Path Connections and Its Applications" (*https://bit.ly/2U1jucF*).

BFS is most commonly used as the basis for other more goal-oriented algorithms. For example, Shortest Path, Connected Components, and Closeness Centrality all use the BFS algorithm. It can also be used to find the shortest path between nodes.

Figure 4-3 shows the order in which we would visit the nodes of our transport graph if we were performing a breadth first search that started from the Dutch city, Den Haag (in English, The Hague). The numbers next to the city name indicate the order in which each node is visited.

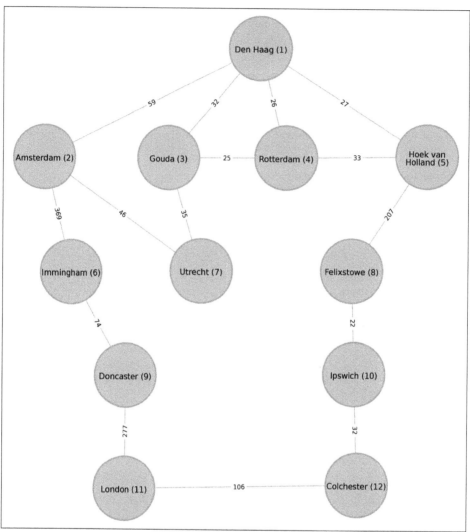

Figure 4-3. Breadth First Search starting from Den Haag. Node numbers indicate the order traversed.

We first visit all of Den Haag's direct neighbors, before visiting their neighbors, and their neighbors' neighbors, until we've run out of relationships to traverse.

Breadth First Search with Apache Spark

Spark's implementation of the Breadth First Search algorithm finds the shortest path between two nodes by the number of relationships (i.e., hops) between them. You can explicitly name your target node or add criteria to be met.

For example, we can use the `bfs` function to find the first medium-sized (by European standards) city that has a population of between 100,000 and 300,000 people. Let's first check which places have a population matching those criteria:

```
(g.vertices
 .filter("population > 100000 and population < 300000")
 .sort("population")
 .show())
```

This is the output we'll see:

id	latitude	longitude	population
Colchester	51.88921	0.90421	104390
Ipswich	52.05917	1.15545	133384

There are only two places matching our criteria, and we'd expect to reach Ipswich first based on a breadth first search.

The following code finds the shortest path from Den Haag to a medium-sized city:

```
from_expr = "id='Den Haag'"
to_expr = "population > 100000 and population < 300000 and id <> 'Den Haag'"
result = g.bfs(from_expr, to_expr)
```

`result` contains columns that describe the nodes and relationships between the two cities. We can run the following code to see the list of columns returned:

```
print(result.columns)
```

This is the output we'll see:

```
['from', 'e0', 'v1', 'e1', 'v2', 'e2', 'to']
```

Columns beginning with e represent relationships (edges) and columns beginning with v represent nodes (vertices). We're only interested in the nodes, so let's filter out any columns that begin with e from the resulting DataFrame:

```
columns = [column for column in result.columns if not column.startswith("e")]
result.select(columns).show()
```

If we run the code in pyspark we'll see this output:

from	v1	v2	to
[Den Haag, 52.078…	[Hoek van Holland…	[Felixstowe, 51.9…	[Ipswich, 52.0591…

As expected, the `bfs` algorithm returns Ipswich! Remember that this function is satisfied when it finds the first match, and as you can see in Figure 4-3, Ipswich is evaluated before Colchester.

Depth First Search

Depth First Search (DFS) is the other fundamental graph traversal algorithm. It starts from a chosen node, picks one of its neighbors, and then traverses as far as it can along that path before backtracking.

DFS was originally invented by French mathematician Charles Pierre Trémaux as a strategy for solving mazes. It provides a useful tool to simulate possible paths for scenario modeling. Figure 4-4 shows the order in which we would visit the nodes of our transport graph if we were performing a DFS that started from Den Haag.

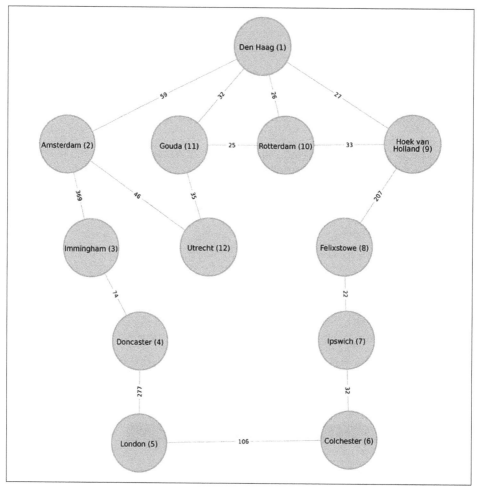

Figure 4-4. Depth First Search starting from Den Haag. Node numbers indicate the order traversed.

Notice how different the node order is compared to BFS. For this DFS, we start by traversing from Den Haag to Amsterdam, and are then able to get to every other node in the graph without needing to backtrack at all!

We can see how search algorithms lay the groundwork for moving through graphs. Now let's look at the pathfinding algorithms that find the cheapest path in terms of the number of hops or weight. Weights can be anything measured, such as time, distance, capacity, or cost.

Two Special Paths/Cycles

There are two special paths in graph analysis that are worth noting. First, an *Eulerian path* is one where every relationship is visited exactly once. Second, a *Hamiltonian path* is one where every node is visited exactly once. A path can be both Eulerian and Hamiltonian, and if you start and finish at the same node it's considered a *cycle* or *tour*. A visual comparison is shown in Figure 4-5.

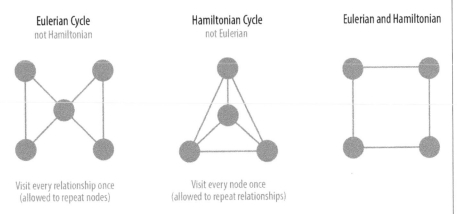

Figure 4-5. *Eulerian and Hamiltonian cycles have a special historical significance.*

The Königsberg bridges problem from Chapter 1 was searching for an Eulerian cycle. It's easy to see how this applies to routing scenarios such as directing snowplows and mail delivery. However, Eulerian paths are also used by other algorithms in processing data in tree structures and are simpler mathematically to study than other cycles.

The Hamiltonian cycle is best known from its relation to the *Traveling Salesman Problem* (TSP), which asks, "What's the shortest possible route for a salesperson to visit each of their assigned cities and return to the origin city?" Although seemingly similar to an Eulerian tour, the TSP is computationally more intensive with approximation alternatives. It's used in a wide variety of planning, logistics, and optimization problems.

Shortest Path

The Shortest Path algorithm calculates the shortest (weighted) path between a pair of nodes. It's useful for user interactions and dynamic workflows because it works in real time.

Pathfinding has a history dating back to the 19th century and is considered to be a classic graph problem. It gained prominence in the early 1950s in the context of alternate routing; that is, finding the second-shortest route if the shortest route is blocked. In 1956, Edsger Dijkstra created the best-known of these algorithms.

Dijkstra's Shortest Path algorithm operates by first finding the lowest-weight relationship from the start node to directly connected nodes. It keeps track of those weights and moves to the "closest" node. It then performs the same calculation, but now as a cumulative total from the start node. The algorithm continues to do this, evaluating a "wave" of cumulative weights and always choosing the lowest weighted cumulative path to advance along, until it reaches the destination node.

 You'll notice in graph analytics the use of the terms *weight*, *cost*, *distance*, and *hop* when describing relationships and paths. "Weight" is the numeric value of a particular property of a relationship. "Cost" is used similarly, but we'll see it more often when considering the total weight of a path.

"Distance" is often used within an algorithm as the name of the relationship property that indicates the cost of traversing between a pair of nodes. It's not required that this be an actual physical measure of distance. "Hop" is commonly used to express the number of relationships between two nodes. You may see some of these terms combined, as in "It's a five-hop distance to London" or "That's the lowest cost for the distance."

When Should I Use Shortest Path?

Use Shortest Path to find optimal routes between a pair of nodes, based on either the number of hops or any weighted relationship value. For example, it can provide real-time answers about degrees of separation, the shortest distance between points, or the least expensive route. You can also use this algorithm to simply explore the connections between particular nodes.

Example use cases include:

- Finding directions between locations. Web-mapping tools such as Google Maps use the Shortest Path algorithm, or a close variant, to provide driving directions.

- Finding the degrees of separation between people in social networks. For example, when you view someone's profile on LinkedIn, it will indicate how many people separate you in the graph, as well as listing your mutual connections.

- Finding the number of degrees of separation between an actor and Kevin Bacon based on the movies they've appeared in (the *Bacon Number*). An example of this can be seen on the Oracle of Bacon website (*https://oracleofbacon.org*). The Erdös Number Project (*https://www.oakland.edu/enp*) provides a similar graph analysis based on collaboration with Paul Erdös, one of the most prolific mathematicians of the twentieth century.

 Dijkstra's algorithm does not support negative weights. The algorithm assumes that adding a relationship to a path can never make a path shorter—an invariant that would be violated with negative weights.

Shortest Path with Neo4j

The Neo4j Graph Data Science library has a built-in procedure that we can use to compute both unweighted and weighted shortest paths. Let's first learn how to compute unweighted shortest paths.

Neo4j's Shortest Path algorithm takes in a config map with the following keys:

startNode
> The node where our shortest path search begins.

endNode
> The node where our shortest path search ends.

nodeProjection
> Enables the mapping of specific kinds of nodes into the in-memory graph. We can declare one or more node labels.

relationshipProjection
> Enables the mapping of relationship types into the in-memory graph. We can declare one or more relationship types along with direction and properties.

relationshipWeightProperty
> The relationship property that indicates the cost of traversing between a pair of nodes. The cost is the number of kilometers between two locations.

To have Neo4j's Shortest Path algorithm ignore weights we won't set the relation shipWeightProperty key. The algorithm will then assume a default weight of 1.0 for each relationship.

The following query computes the unweighted shortest path from Amsterdam to London:

```
MATCH (source:Place {id: "Amsterdam"}),
      (destination:Place {id: "London"})

CALL gds.alpha.shortestPath.stream({
  startNode: source,
  endNode: destination,
  nodeProjection: "*",
  relationshipProjection: {
    all: {
      type: "*",
      orientation:  "UNDIRECTED"
    }
  }
})
YIELD nodeId, cost
RETURN gds.util.asNode(nodeId).id AS place, cost;
```

In this query we are passing `nodeProjection: "*"`, which means that all node labels should be considered. The `relationshipProjection` is a bit more complicated. We're using the advanced configuration mode, which enables a more flexible definition of the relationship types to consider during the traversal. Let's break down the values that we passed in:

`type: "*"`
  All relationship types should be considered.

`orientation: "UNDIRECTED"`
  Each relationship in the underlying graph is projected in both directions.

 More detailed documentation about node and relationship projections can be found in the Native Projection chapter (*https://neo4j.com/docs/graph-data-science/current/management-ops/native-projection/*) of the Graph Data Science user manual.

This query returns the following output:

place	cost
Amsterdam	0.0
Immingham	1.0
Doncaster	2.0
London	3.0

Here, the cost is the cumulative total for relationships (or hops). This is the same path as we see using Breadth First Search in Spark.

We could even work out the total distance of following this path by writing a bit of postprocessing Cypher. The following procedure calculates the shortest unweighted path and then works out what the actual cost of that path would be:

```
MATCH (source:Place {id: "Amsterdam"}),
      (destination:Place {id: "London"})

CALL gds.alpha.shortestPath.stream({
  startNode: source,
  endNode: destination,
  nodeProjection: "*",
  relationshipProjection: {
    all: {
      type: "*",
      orientation:  "UNDIRECTED"
    }
  }
})
YIELD nodeId, cost

WITH collect(gds.util.asNode(nodeId)) AS path
UNWIND range(0, size(path)-1) AS index
WITH path[index] AS current, path[index+1] AS next
WITH current, next, [(current)-[r:EROAD]-(next) | r.distance][0] AS distance

WITH collect({current: current, next:next, distance: distance}) AS stops
UNWIND range(0, size(stops)-1) AS index
WITH stops[index] AS location, stops, index
RETURN location.current.id AS place,
       reduce(acc=0.0,
              distance in [stop in stops[0..index] | stop.distance] |
              acc + distance) AS cost;
```

If the previous code feels a bit unwieldy, notice that the tricky part is figuring out how to massage the data to include the cost over the whole journey. This is helpful to keep in mind when we need the cumulative path cost.

The query returns the following result:

place	cost
Amsterdam	0.0
Immingham	369.0
Doncaster	443.0
London	720.0

Figure 4-6 shows the unweighted shortest path from Amsterdam to London, routing us through the fewest number of cities. It has a total cost of 720 km.

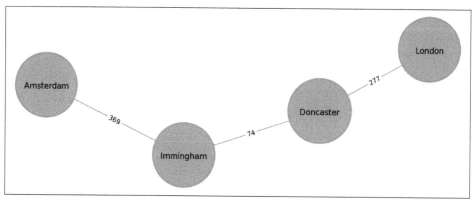

Figure 4-6. The unweighted shortest path between Amsterdam and London

Choosing a route with the fewest number of nodes visited might be very useful in situations such as subway systems, where less stops are highly desirable. However, in a driving scenario, we're probably more interested in the total cost using the shortest weighted path.

Shortest Path (Weighted) with Neo4j

We can execute the Weighted Shortest Path algorithm to find the shortest path between Amsterdam and London like this:

```
MATCH (source:Place {id: "Amsterdam"}),
      (destination:Place {id: "London"})

CALL gds.alpha.shortestPath.stream({
  startNode: source,
  endNode: destination,
  nodeProjection: "*",
  relationshipProjection: {
    all: {
      type: "*",
      properties: "distance",
      orientation:  "UNDIRECTED"
    }
  },
  relationshipWeightProperty: "distance"
})
YIELD nodeId, cost
RETURN gds.util.asNode(nodeId).id AS place, cost;
```

We are now passing the optional `relationshipWeightProperty`, which is the name of the relationship property that indicates the cost of traversing between a pair of nodes.

The cost is the number of kilometers between two locations. The query returns the following result:

place	cost
Amsterdam	0.0
Den Haag	59.0
Hoek van Holland	86.0
Felixstowe	293.0
Ipswich	315.0
Colchester	347.0
London	453.0

The quickest route takes us via Den Haag, Hoek van Holland, Felixstowe, Ipswich, and Colchester! The cost shown is the cumulative total as we progress through the cities. First we go from Amsterdam to Den Haag, at a cost of 59. Then we go from Den Haag to Hoek van Holland, at a cumulative cost of 86—and so on. Finally, we arrive in London, from Colchester, for a total cost of 453 km.

Remember that the unweighted shortest path had a total cost of 720 km, so we've been able to save 267 km by taking weights into account when computing the shortest path.

Shortest Path (Weighted) with Apache Spark

In the Breadth First Search with Apache Spark section we learned how to find the shortest path between two nodes. That shortest path was based on hops and therefore isn't the same as the shortest *weighted* path, which would tell us the shortest total distance between cities.

If we want to find the shortest weighted path (in this case, distance) we need to use the `cost` property, which is used for various types of weighting. This option is not available out of the box with GraphFrames, so we need to write our own version of Weighted Shortest Path using its `aggregateMessages` framework (*https://bit.ly/ 2JCFBRJ*). Most of our algorithm examples for Spark use the simpler process of calling on algorithms from the library, but we have the option of writing our own functions. More information on `aggregateMessages` can be found in the "Message passing via AggregateMessages" (*http://bit.ly/2Wo6Hxg*) section of the GraphFrames user guide.

When available, we recommend leveraging preexisting, tested libraries. Writing our own functions, especially for more complicated algorithms, requires a deeper understanding of our data and calculations.

The following example should be treated as a reference implementation, and would need to be optimized before running on a larger dataset. Those that aren't interested in writing their own functions can skip this example.

Before we create our function, we'll import some libraries that we'll use:

```
from graphframes.lib import AggregateMessages as AM
from pyspark.sql import functions as F
```

The `aggregateMessages` module is part of the GraphFrames library and contains some useful helper functions.

Now let's write our function. We first create a user-defined function that we'll use to build the paths between our source and destination:

```
add_path_udf = F.udf(lambda path, id: path + [id], ArrayType(StringType()))
```

And now for the main function, which calculates the shortest path starting from an origin and returns as soon as the destination has been visited:

```
def shortest_path(g, origin, destination, column_name="cost"):
    if g.vertices.filter(g.vertices.id == destination).count() == 0:
        return (spark.createDataFrame(sc.emptyRDD(), g.vertices.schema)
                .withColumn("path", F.array()))

    vertices = (g.vertices.withColumn("visited", F.lit(False))
                .withColumn("distance", F.when(g.vertices["id"] == origin, 0)
                            .otherwise(float("inf")))
                .withColumn("path", F.array()))
    cached_vertices = AM.getCachedDataFrame(vertices)
    g2 = GraphFrame(cached_vertices, g.edges)

    while g2.vertices.filter('visited == False').first():
        current_node_id = g2.vertices.filter('visited == False').sort
                                        ("distance").first().id

        msg_distance = AM.edge[column_name] + AM.src['distance']
        msg_path = add_path_udf(AM.src["path"], AM.src["id"])
        msg_for_dst = F.when(AM.src['id'] == current_node_id,
                             F.struct(msg_distance, msg_path))
        new_distances = g2.aggregateMessages(F.min(AM.msg).alias("aggMess"),
                                             sendToDst=msg_for_dst)

        new_visited_col = F.when(
            g2.vertices.visited | (g2.vertices.id == current_node_id),
                                        True).otherwise(False)
```

```
new_distance_col = F.when(new_distances["aggMess"].isNotNull() &
                          (new_distances.aggMess["col1"]
                          < g2.vertices.distance),
                          new_distances.aggMess["col1"])
                   .otherwise(g2.vertices.distance)
new_path_col = F.when(new_distances["aggMess"].isNotNull() &
                      (new_distances.aggMess["col1"]
                      < g2.vertices.distance), new_distances.aggMess["col2"]
                      .cast("array<string>")).otherwise(g2.vertices.path)

new_vertices = (g2.vertices.join(new_distances, on="id",
                                 how="left_outer")
                .drop(new_distances["id"])
                .withColumn("visited", new_visited_col)
                .withColumn("newDistance", new_distance_col)
                .withColumn("newPath", new_path_col)
                .drop("aggMess", "distance", "path")
                .withColumnRenamed('newDistance', 'distance')
                .withColumnRenamed('newPath', 'path'))
cached_new_vertices = AM.getCachedDataFrame(new_vertices)
g2 = GraphFrame(cached_new_vertices, g2.edges)
if g2.vertices.filter(g2.vertices.id == destination).first().visited:
    return (g2.vertices.filter(g2.vertices.id == destination)
            .withColumn("newPath", add_path_udf("path", "id"))
            .drop("visited", "path")
            .withColumnRenamed("newPath", "path"))
return (spark.createDataFrame(sc.emptyRDD(), g.vertices.schema)
        .withColumn("path", F.array()))
```

 If we store references to any DataFrames in our functions, we need
to cache them using the AM.getCachedDataFrame function or we'll
encounter a memory leak during execution. In the shortest_path
function we use this function to cache the vertices and new_verti
ces DataFrames.

If we wanted to find the shortest path between Amsterdam and Colchester we could
call that function like so:

```
result = shortest_path(g, "Amsterdam", "Colchester", "cost")
result.select("id", "distance", "path").show(truncate=False)
```

which would return the following result:

id	distance	path
Colchester	347.0	[Amsterdam, Den Haag, Hoek van Holland, Felixstowe, Ipswich, Colchester]

The total distance of the shortest path between Amsterdam and Colchester is 347 km
and takes us via Den Haag, Hoek van Holland, Felixstowe, and Ipswich. By contrast,
the shortest path in terms of number of relationships between the locations, which we

worked out with the Breadth First Search algorithm (refer back to Figure 4-4), would take us via Immingham, Doncaster, and London.

Shortest Path Variation: A*

The A* Shortest Path algorithm improves on Dijkstra's by finding shortest paths more quickly. It does this by allowing the inclusion of extra information that the algorithm can use, as part of a heuristic function, when determining which paths to explore next.

The algorithm was invented by Peter Hart, Nils Nilsson, and Bertram Raphael and described in their 1968 paper "A Formal Basis for the Heuristic Determination of Minimum Cost Paths" (*https://bit.ly/2JAaV3s*).

The A* algorithm operates by determining which of its partial paths to expand at each iteration of its main loop. It does so based on an estimate of the cost (heuristic) still left to reach the goal node.

 Be thoughtful in the heuristic employed to estimate path costs. Underestimating path costs may unnecessarily include some paths that could have been eliminated, but the results will still be accurate. However, if the heuristic overestimates path costs, it may skip over actual shorter paths (incorrectly estimated to be longer) that should have been evaluated, which can lead to inaccurate results.

A* selects the path that minimizes the following function:

`f(n) = g(n) + h(n)`

where:

- g(n) is the cost of the path from the starting point to node n.
- h(n) is the estimated cost of the path from node n to the destination node, as computed by a heuristic.

 In Neo4j's implementation, geospatial distance is used as the heuristic. In our example transportation dataset we use the latitude and longitude of each location as part of the heuristic function.

A* with Neo4j

Neo4j's A* algorithm takes in a config map with the following keys:

startNode
: The node where our shortest path search begins.

endNode
: The node where our shortest path search ends.

nodeProjection
: Enables the mapping of specific kinds of nodes into the in-memory graph. We can declare one or more node labels.

relationshipProjection
: Enables the mapping of relationship types into the in-memory graph. We can declare one or more relationship types along with direction and properties.

relationshipWeightProperty
: The relationship property that indicates the cost of traversing between a pair of nodes. The cost is the number of kilometers between two locations.

propertyKeyLat
: The name of the node property used to represent the latitude of each node as part of the geospatial heuristic calculation.

propertyKeyLon
: The name of the node property used to represent the longitude of each node as part of the geospatial heuristic calculation.

The following query executes the A* algorithm to find the shortest path between Den Haag and London:

```
MATCH (source:Place {id: "Den Haag"}),
      (destination:Place {id: "London"})
CALL gds.alpha.shortestPath.astar.stream({
  startNode: source,
  endNode: destination,
  nodeProjection: "*",
  relationshipProjection: {
    all: {
      type: "*",
      properties: "distance",
      orientation:  "UNDIRECTED"
    }
  },
  relationshipWeightProperty: "distance",
  propertyKeyLat: "latitude",
  propertyKeyLon: "longitude"
})
YIELD nodeId, cost
RETURN gds.util.asNode(nodeId).id AS place, cost;
```

Running this procedure gives the following result:

place	cost
Den Haag	0.0
Hoek van Holland	27.0
Felixstowe	234.0
Ipswich	256.0
Colchester	288.0
London	394.0

We'd get the same result using the Shortest Path algorithm, but on more complex datasets the A* algorithm will be faster as it evaluates fewer paths.

Shortest Path Variation: Yen's k-Shortest Paths

Yen's k-Shortest Paths algorithm is similar to the Shortest Path algorithm, but rather than finding just the shortest path between two pairs of nodes, it also calculates the second shortest path, third shortest path, and so on up to k-1 deviations of shortest paths.

Jin Y. Yen invented the algorithm in 1971 and described it in "Finding the K Shortest Loopless Paths in a Network" (*https://bit.ly/2HS0eXB*). This algorithm is useful for getting alternative paths when finding the absolute shortest path isn't our only goal. It can be particularly helpful when we need more than one backup plan!

Yen's with Neo4j

The Yen's algorithm takes in a config map with the following keys:

startNode
> The node where our shortest path search begins.

endNode
> The node where our shortest path search ends.

nodeProjection
> Enables the mapping of specific kinds of nodes into the in-memory graph. We can declare one or more node labels.

relationshipProjection
> Enables the mapping of relationship types into the in-memory graph. We can declare one or more relationship types along with direction and properties.

relationshipWeightProperty
> The relationship property that indicates the cost of traversing between a pair of nodes. The cost is the number of kilometers between two locations.

k

The maximum number of shortest paths to find.

The following query executes Yen's algorithm to find the shortest paths between Gouda and Felixstowe:

```
MATCH (start:Place {id:"Gouda"}),
      (end:Place {id:"Felixstowe"})

CALL gds.alpha.kShortestPaths.stream({
  startNode: start,
  endNode: end,
  nodeProjection: "*",
  relationshipProjection: {
    all: {
      type: "*",
      properties: "distance",
      orientation:  "UNDIRECTED"
    }
  },
  relationshipWeightProperty: "distance",
  k: 5
})

YIELD index, sourceNodeId, targetNodeId, nodeIds, costs, path
RETURN index,
       [node in gds.util.asNodes(nodeIds[1..-1]) | node.id] AS via,
       reduce(acc=0.0, cost in costs | acc + cost) AS totalCost;
```

After we get back the shortest paths, we look up the associated node for each node ID using the `gds.util.asNodes` function, and then filter out the start and end nodes from the resulting collection. We also calculate the total cost for each path by summing the returned costs.

Running this procedure gives the following result:

index	via	totalCost
0	[Rotterdam, Hoek van Holland]	265.0
1	[Den Haag, Hoek van Holland]	266.0
2	[Rotterdam, Den Haag, Hoek van Holland]	285.0
3	[Den Haag, Rotterdam, Hoek van Holland]	298.0
4	[Utrecht, Amsterdam, Den Haag, Hoek van Holland]	374.0

Figure 4-7 shows the shortest path between Gouda and Felixstowe.

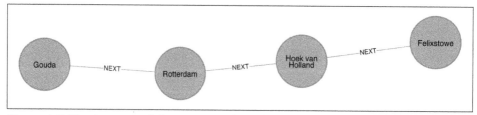

Figure 4-7. The shortest path between Gouda and Felixstowe

The shortest path in Figure 4-7 is interesting in comparison to the results ordered by total cost. It illustrates that sometimes you may want to consider several shortest paths or other parameters. In this example, the second-shortest route is only 1 km longer than the shortest one. If we prefer the scenery, we might choose the slightly longer route.

All Pairs Shortest Path

The All Pairs Shortest Path (APSP) algorithm calculates the shortest (weighted) path between all pairs of nodes. It's more efficient than running the Single Source Shortest Path algorithm for every pair of nodes in the graph.

APSP optimizes operations by keeping track of the distances calculated so far and running on nodes in parallel. Those known distances can then be reused when calculating the shortest path to an unseen node. You can follow the example in the next section to get a better understanding of how the algorithm works.

 Some pairs of nodes might not be reachable from each other, which means that there is no shortest path between these nodes. The algorithm doesn't return distances for these pairs of nodes.

A Closer Look at All Pairs Shortest Path

The calculation for APSP is easiest to understand when you follow a sequence of operations. The diagram in Figure 4-8 walks through the steps for node A.

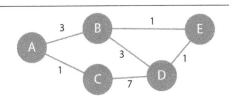

All nodes start with a ∞ distance and then the start node is set to a 0 distance		Each Step Keeps or Updates to the Lowest Value Calculated so Far Only steps for node A to all nodes shown				
		1st from A	2nd from A to C to Next	3rd from A to B to Next	4th from A to E to Next	5th from A to D to Next
A	∞	0	0	0	0	0	0
B	∞	∞	3	3	3	3	3
C	∞	∞	1	1	1	1	1
D	∞	∞	∞	8	6	5	5
E	∞	∞	∞	∞	4	4	4

Figure 4-8. The steps to calculate the shortest path from node A to all other nodes, with updates shaded.

Initially the algorithm assumes an infinite distance to all nodes. When a start node is selected, then the distance to that node is set to 0. The calculation then proceeds as follows:

1. From start node A we evaluate the cost of moving to the nodes we can reach and update those values. Looking for the smallest value, we have a choice of B (cost of 3) or C (cost of 1). C is selected for the next phase of traversal.

2. Now from node C, the algorithm updates the cumulative distances from A to nodes that can be reached directly from C. Values are only updated when a lower cost has been found:

   A=0, B=3, C=1, D=8, E=∞

3. Then B is selected as the next closest node that hasn't already been visited. It has relationships to nodes A, D, and E. The algorithm works out the distance to those nodes by summing the distance from A to B with the distance from B to each of those nodes. Note that the lowest cost from the start node A to the current node is always preserved as a sunk cost. The distance (d) calculation results:

   d(A,A) = d(A,B) + d(B,A) = 3 + 3 = 6
   d(A,D) = d(A,B) + d(B,D) = 3 + 3 = 6
   d(A,E) = d(A,B) + d(B,E) = 3 + 1 = 4

- In this step the distance from node A to B and back to A, shown as d(A,A) = 6, is greater than the shortest distance already computed (0), so its value is not updated.

- The distances for nodes D (6) and E (4) are less than the previously calculated distances, so their values are updated.

4. E is selected next. Only the cumulative total for reaching D (5) is now lower, and therefore it is the only one updated.

5. When D is finally evaluated, there are no new minimum path weights; nothing is updated, and the algorithm terminates.

Even though the All Pairs Shortest Path algorithm is optimized to run calculations in parallel for each node, this can still add up for a very large graph. Consider using a subgraph if you only need to evaluate paths between a subcategory of nodes.

When Should I Use All Pairs Shortest Path?

All Pairs Shortest Path is commonly used for understanding alternate routing when the shortest route is blocked or becomes suboptimal. For example, this algorithm is used in logical route planning to ensure the best multiple paths for diversity routing. Use All Pairs Shortest Path when you need to consider all possible routes between all or most of your nodes.

Example use cases include:

- Optimizing the location of urban facilities and the distribution of goods. One example of this is determining the traffic load expected on different segments of a transportation grid. For more information, see R. C. Larson and A. R. Odoni's book, *Urban Operations Research* (Prentice-Hall).

- Finding a network with maximum bandwidth and minimal latency as part of a data center design algorithm. There are more details about this approach in the paper "REWIRE: An Optimization-Based Framework for Data Center Network Design" (*https://bit.ly/2HTbhzY*), by A. R. Curtis et al.

All Pairs Shortest Path with Apache Spark

Spark's shortestPaths function is designed for finding the shortest paths from all nodes to a set of nodes called *landmarks*. If we wanted to find the shortest path from every location to Colchester, Immingham, and Hoek van Holland, we would write the following query:

```
result = g.shortestPaths(["Colchester", "Immingham", "Hoek van Holland"])
result.sort(["id"]).select("id", "distances").show(truncate=False)
```

If we run that code in pyspark we'll see this output:

id	distances
Amsterdam	[Immingham → 1, Hoek van Holland → 2, Colchester → 4]
Colchester	[Colchester → 0, Hoek van Holland → 3, Immingham → 3]
Den Haag	[Hoek van Holland → 1, Immingham → 2, Colchester → 4]
Doncaster	[Immingham → 1, Colchester → 2, Hoek van Holland → 4]
Felixstowe	[Hoek van Holland → 1, Colchester → 2, Immingham → 4]
Gouda	[Hoek van Holland → 2, Immingham → 3, Colchester → 5]
Hoek van Holland	[Hoek van Holland → 0, Immingham → 3, Colchester → 3]
Immingham	[Immingham → 0, Colchester → 3, Hoek van Holland → 3]
Ipswich	[Colchester → 1, Hoek van Holland → 2, Immingham → 4]
London	[Colchester → 1, Immingham → 2, Hoek van Holland → 4]
Rotterdam	[Hoek van Holland → 1, Immingham → 3, Colchester → 4]
Utrecht	[Immingham → 2, Hoek van Holland → 3, Colchester → 5]

The number next to each location in the distances column is the number of relationships (roads) between cities we need to traverse to get there from the source node. In our example, Colchester is one of our destination cities and you can see it has 0 nodes to traverse to get to itself but 3 hops to make from Immingham and Hoek van Holland. If we were planning a trip, we could use this information to help maximize our time at our chosen destinations.

All Pairs Shortest Path with Neo4j

Neo4j has a parallel implementation of the All Pairs Shortest Path algorithm, which returns the distance between every pair of nodes.

The All Pairs Shortest Path algorithm takes in a config map with the following keys:

nodeProjection
> Enables the mapping of specific kinds of nodes into the in-memory graph. We can declare one or more node labels.

relationshipProjection
> Enables the mapping of relationship types into the in-memory graph. We can declare one or more relationship types along with direction and properties.

relationshipWeightProperty
> The relationship property that indicates the cost of traversing between a pair of nodes. The cost is the number of kilometers between two locations.

If we don't set `relationshipWeightProperty` then the algorithm will calculate the unweighted shortest paths between all pairs of nodes.

The following query does this:

```
CALL gds.alpha.allShortestPaths.stream({
  nodeProjection: "*",
  relationshipProjection: {
    all: {
      type: "*",
      properties: "distance",
      orientation: "UNDIRECTED"
    }
  }
})
YIELD sourceNodeId, targetNodeId, distance
WHERE sourceNodeId < targetNodeId
RETURN gds.util.asNode(sourceNodeId).id AS source,
       gds.util.asNode(targetNodeId).id AS target,
       distance
ORDER BY distance DESC
LIMIT 10;
```

This algorithm returns the shortest path between every pair of nodes twice—once with each of the nodes as the source node. This would be helpful if you were evaluating a directed graph of one-way streets. However, we don't need to see each path twice, so we filter the results to only keep one of them by using the `sourceNodeId < targetNodeId` predicate.

The query returns the following results:

source	target	distance
Colchester	Utrecht	5.0
London	Rotterdam	5.0
London	Gouda	5.0
Ipswich	Utrecht	5.0
Colchester	Gouda	5.0
Colchester	Den Haag	4.0
London	Utrecht	4.0
London	Den Haag	4.0
Colchester	Amsterdam	4.0
Ipswich	Gouda	4.0

This output shows the 10 pairs of locations that have the most relationships between them because we asked for results in descending order (DESC).

If we want to calculate the shortest weighted paths, we should set `relationship WeightProperty` to the property name that contains the `cost` to be used in the shortest path calculation. This property will then be evaluated to work out the shortest weighted path between each pair of nodes.

The following query does this:

```
CALL gds.alpha.allShortestPaths.stream({
  nodeProjection: "*",
  relationshipProjection: {
    all: {
      type: "*",
      properties: "distance",
      orientation: "UNDIRECTED"
    }
  },
  relationshipWeightProperty: "distance"
})
YIELD sourceNodeId, targetNodeId, distance
WHERE sourceNodeId < targetNodeId
RETURN gds.util.asNode(sourceNodeId).id AS source,
       gds.util.asNode(targetNodeId).id AS target,
       distance
ORDER BY distance DESC
LIMIT 10;
```

The query returns the following result:

source	target	distance
Doncaster	Hoek van Holland	529.0
Rotterdam	Doncaster	528.0
Gouda	Doncaster	524.0
Felixstowe	Immingham	511.0
Den Haag	Doncaster	502.0
Ipswich	Immingham	489.0
Utrecht	Doncaster	489.0
London	Utrecht	460.0
Colchester	Immingham	457.0
Immingham	Hoek van Holland	455.0

Now we're seeing the 10 pairs of locations furthest from each other in terms of the total distance between them. Notice that Doncaster shows up frequently along with several cities in the Netherlands. It looks like it would be a long drive if we wanted to take a road trip between those areas.

Single Source Shortest Path

The Single Source Shortest Path (SSSP) algorithm, which came into prominence at around the same time as Dijkstra's Shortest Path algorithm, acts as an implementation for both problems.

The SSSP algorithm calculates the shortest (weighted) path from a root node to all other nodes in the graph, as demonstrated in Figure 4-9.

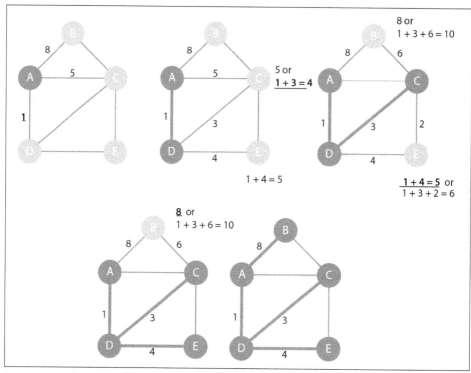

Figure 4-9. The steps of the Single Source Shortest Path algorithm

It proceeds as follows:

1. It begins with a root node from which all paths will be measured. In Figure 4-9 we've selected node A as the root.

2. The relationship with the smallest weight coming from that root node is selected and added to the tree, along with its connected node. In this case, that's d(A,D)=1.

3. The next relationship with the smallest cumulative weight from our root node to any unvisited node is selected and added to the tree in the same way. Our choices

in Figure 4-9 are d(A,B)=8, d(A,C)=5 directly or 4 via A-D-C, and d(A,E)=5. So, the route via A-D-C is chosen and C is added to our tree.

4. The process continues until there are no more nodes to add and we have our single source shortest path.

When Should I Use Single Source Shortest Path?

Use Single Source Shortest Path when you need to evaluate the optimal route from a fixed start point to all other individual nodes. Because the route is chosen based on the total path weight from the root, it's useful for finding the best path to each node, but not necessarily when all nodes need to be visited in a single trip.

For example, SSSP is helpful for identifying the main routes to use for emergency services where you don't visit every location on each incident, but not for finding a single route for garbage collection where you need to visit each house in one trip. (In the latter case, you'd use the Minimum Spanning Tree algorithm, covered later.)

Example use cases include:

- Detecting changes in topology, such as link failures, and suggesting a new routing structure in seconds (*https://bit.ly/2HL7ndd*)
- Using Dijkstra as an IP routing protocol for use in autonomous systems such as a local area network (LAN) (*https://bit.ly/2HUsAAr*)

Single Source Shortest Path with Apache Spark

We can adapt the shortest_path function that we wrote to calculate the shortest path between two locations to instead return us the shortest path from one location to all others. Note that we're using Spark's aggregateMessages framework again to customize our function.

We'll first import the same libraries as before:

```
from graphframes.lib import AggregateMessages as AM
from pyspark.sql import functions as F
```

And we'll use the same user-defined function to construct paths:

```
add_path_udf = F.udf(lambda path, id: path + [id], ArrayType(StringType()))
```

Now for the main function, which calculates the shortest path starting from an origin:

```
def sssp(g, origin, column_name="cost"):
    vertices = g.vertices \
        .withColumn("visited", F.lit(False)) \
        .withColumn("distance",
            F.when(g.vertices["id"] == origin, 0).otherwise(float("inf"))) \
        .withColumn("path", F.array())
```

```
        cached_vertices = AM.getCachedDataFrame(vertices)
        g2 = GraphFrame(cached_vertices, g.edges)

        while g2.vertices.filter('visited == False').first():
            current_node_id = g2.vertices.filter('visited == False') \
                              .sort("distance").first().id

            msg_distance = AM.edge[column_name] + AM.src['distance']
            msg_path = add_path_udf(AM.src["path"], AM.src["id"])
            msg_for_dst = F.when(AM.src['id'] == current_node_id,
                          F.struct(msg_distance, msg_path))
            new_distances = g2.aggregateMessages(
                F.min(AM.msg).alias("aggMess"), sendToDst=msg_for_dst)

            new_visited_col = F.when(
                g2.vertices.visited | (g2.vertices.id == current_node_id),
                                True).otherwise(False)
            new_distance_col = F.when(new_distances["aggMess"].isNotNull() &
                                      (new_distances.aggMess["col1"] <
                                      g2.vertices.distance),
                                      new_distances.aggMess["col1"]) \
                                      .otherwise(g2.vertices.distance)
            new_path_col = F.when(new_distances["aggMess"].isNotNull() &
                                  (new_distances.aggMess["col1"] <
                                  g2.vertices.distance),
                                  new_distances.aggMess["col2"]
                                  .cast("array<string>")) \
                                  .otherwise(g2.vertices.path)

            new_vertices = g2.vertices.join(new_distances, on="id",
                                            how="left_outer") \
                .drop(new_distances["id"]) \
                .withColumn("visited", new_visited_col) \
                .withColumn("newDistance", new_distance_col) \
                .withColumn("newPath", new_path_col) \
                .drop("aggMess", "distance", "path") \
                .withColumnRenamed('newDistance', 'distance') \
                .withColumnRenamed('newPath', 'path')
            cached_new_vertices = AM.getCachedDataFrame(new_vertices)
            g2 = GraphFrame(cached_new_vertices, g2.edges)

        return g2.vertices \
                .withColumn("newPath", add_path_udf("path", "id")) \
                .drop("visited", "path") \
                .withColumnRenamed("newPath", "path")
```

If we want to find the shortest path from Amsterdam to all other locations we can call the function like this:

```
via_udf = F.udf(lambda path: path[1:-1], ArrayType(StringType()))

result = sssp(g, "Amsterdam", "cost")
(result
```

```
.withColumn("via", via_udf("path"))
.select("id", "distance", "via")
.sort("distance")
.show(truncate=False))
```

We define another user-defined function to filter out the start and end nodes from the resulting path. If we run that code we'll see the following output:

id	distance	via
Amsterdam	0.0	[]
Utrecht	46.0	[]
Den Haag	59.0	[]
Gouda	81.0	[Utrecht]
Rotterdam	85.0	[Den Haag]
Hoek van Holland	86.0	[Den Haag]
Felixstowe	293.0	[Den Haag, Hoek van Holland]
Ipswich	315.0	[Den Haag, Hoek van Holland, Felixstowe]
Colchester	347.0	[Den Haag, Hoek van Holland, Felixstowe, Ipswich]
Immingham	369.0	[]
Doncaster	443.0	[Immingham]
London	453.0	[Den Haag, Hoek van Holland, Felixstowe, Ipswich, Colchester]

In these results we see the physical distances in kilometers from the root node, Amsterdam, to all other cities in the graph, ordered by shortest distance.

Single Source Shortest Path with Neo4j

Neo4j implements a variation of SSSP, called the Delta-Stepping algorithm (*https://bit.ly/2UaCHrw*) that divides Dijkstra's algorithm into a number of phases that can be executed in parallel.

The Single Source Shortest Path algorithm takes in a config map with the following keys:

startNode
> The node where our shortest path search begins.

nodeProjection
> Enables the mapping of specific kinds of nodes into the in-memory graph. We can declare one or more node labels.

relationshipProjection
> Enables the mapping of relationship types into the in-memory graph. We can declare one or more relationship types along with direction and properties.

`relationshipWeightProperty`
> The relationship property that indicates the cost of traversing between a pair of nodes. The cost is the number of kilometers between two locations.

`delta`
> The grade of concurrency to use

The following query executes the Delta-Stepping algorithm:

```
MATCH (n:Place {id:"London"})
CALL gds.alpha.shortestPath.deltaStepping.stream({
  startNode: n,
  nodeProjection: "*",
  relationshipProjection: {
    all: {
      type: "*",
      properties: "distance",
      orientation:  "UNDIRECTED"
    }
  },
  relationshipWeightProperty: "distance",
  delta: 1.0
})
YIELD nodeId, distance
WHERE gds.util.isFinite(distance)
RETURN gds.util.asNode(nodeId).id AS destination, distance
ORDER BY distance;
```

The query returns the following output:

destination	distance
London	0.0
Colchester	106.0
Ipswich	138.0
Felixstowe	160.0
Doncaster	277.0
Immingham	351.0
Hoek van Holland	367.0
Den Haag	394.0
Rotterdam	400.0
Gouda	425.0
Amsterdam	453.0
Utrecht	460.0

In these results we see the physical distances in kilometers from the root node, London, to all other cities in the graph, ordered by shortest distance.

Minimum Spanning Tree

The Minimum (Weight) Spanning Tree algorithm starts from a given node and finds all its reachable nodes and the set of relationships that connect the nodes together with the minimum possible weight. It traverses to the next unvisited node with the lowest weight from any visited node, avoiding cycles.

The first known Minimum Weight Spanning Tree algorithm was developed by the Czech scientist Otakar Borůvka in 1926. Prim's algorithm, invented in 1957, is the simplest and best known.

Prim's algorithm is similar to Dijkstra's Shortest Path algorithm, but rather than minimizing the total length of a path ending at each relationship, it minimizes the length of each relationship individually. Unlike Dijkstra's algorithm, it tolerates negative-weight relationships.

The Minimum Spanning Tree algorithm operates as demonstrated in Figure 4-10.

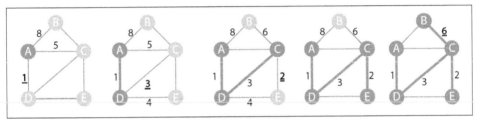

Figure 4-10. The steps of the Minimum Spanning Tree algorithm

The steps are as follows:

1. It begins with a tree containing only one node. In Figure 4-10 we start with node A.

2. The relationship with smallest weight coming from that node is selected and added to the tree (along with its connected node). In this case, A-D.

3. This process is repeated, always choosing the minimal-weight relationship that joins any node not already in the tree. If you compare our example here to the SSSP example in Figure 4-9 you'll notice that in the fourth graph the paths become different. This is because SSSP evaluates the shortest path based on cumulative totals from the root, whereas Minimum Spanning Tree only looks at the cost of the next step.

4. When there are no more nodes to add, the tree is a minimum spanning tree.

There are also variants of this algorithm that find the maximum-weight spanning tree (highest-cost tree) and the *k*-spanning tree (tree size limited).

When Should I Use Minimum Spanning Tree?

Use Minimum Spanning Tree when you need the best route to visit all nodes. Because the route is chosen based on the cost of each next step, it's useful when you must visit all nodes in a single walk. (Review the previous section on "Single Source Shortest Path" on page 68 if you don't need a path for a single trip.)

You can use this algorithm for optimizing paths for connected systems like water pipes and circuit design. It's also employed to approximate some problems with unknown compute times, such as the Traveling Salesman Problem and certain types of rounding problems. Although it may not always find the absolute optimal solution, this algorithm makes potentially complicated and compute-intensive analysis much more approachable.

Example use cases include:

- Minimizing the travel cost of exploring a country. "An Application of Minimum Spanning Trees to Travel Planning" (*https://bit.ly/2CQBs6Q*) describes how the algorithm analyzed airline and sea connections to do this.

- Visualizing correlations between currency returns. This is described in "Minimum Spanning Tree Application in the Currency Market" (*https://bit.ly/2HFbGGG*).

- Tracing the history of infection transmission in an outbreak. For more information, see "Use of the Minimum Spanning Tree Model for Molecular Epidemiological Investigation of a Nosocomial Outbreak of Hepatitis C Virus Infection" (*https://bit.ly/2U7SR4Y*).

> The Minimum Spanning Tree algorithm only gives meaningful results when run on a graph where the relationships have different weights. If the graph has no weights, or all relationships have the same weight, then any spanning tree is a minimum spanning tree.

Minimum Spanning Tree with Neo4j

Let's see the Minimum Spanning Tree algorithm in action. The Minimum Spanning Tree algorithm takes in a config map with the following keys:

startNodeId
:   The id of the node where our shortest path search begins.

nodeProjection
:   Enables the mapping of specific kinds of nodes into the in-memory graph. We can declare one or more node labels.

`relationshipProjection`

Enables the mapping of relationship types into the in-memory graph. We can declare one or more relationship types along with direction and properties.

`relationshipWeightProperty`

The relationship property that indicates the cost of traversing between a pair of nodes. The cost is the number of kilometers between two locations.

`writeProperty`

The name of the relationship type written back as a result

`weightWriteProperty`

The name of the weight property on the `writeProperty` relationship type written back

The following query finds a spanning tree starting from Amsterdam:

```
MATCH (n:Place {id:"Amsterdam"})
CALL gds.alpha.spanningTree.minimum.write({
  startNodeId: id(n),
  nodeProjection: "*",
  relationshipProjection: {
    EROAD: {
      type: "EROAD",
      properties: "distance",
      orientation:  "UNDIRECTED"
    }
  },
  relationshipWeightProperty: "distance",
  writeProperty: 'MINST',
  weightWriteProperty: 'cost'
})
YIELD createMillis, computeMillis, writeMillis, effectiveNodeCount
RETURN createMillis, computeMillis, writeMillis, effectiveNodeCount;
```

The parameters passed to this algorithm are:

`Place`

The node labels to consider when computing the spanning tree

`EROAD`

The relationship types to consider when computing the spanning tree

`distance`

The name of the relationship property that indicates the cost of traversing between a pair of nodes

`id(n)`

The internal node id of the node from which the spanning tree should begin

This query stores its results in the graph. If we want to return the minimum weight spanning tree we can run the following query:

```
MATCH path = (n:Place {id:"Amsterdam"})-[:MINST*]-()
WITH relationships(path) AS rels
UNWIND rels AS rel
WITH DISTINCT rel AS rel
RETURN startNode(rel).id AS source,
       endNode(rel).id AS destination,
       rel.cost AS cost;
```

And this is the output of the query:

source	destination	cost
Amsterdam	Utrecht	46.0
Utrecht	Gouda	35.0
Gouda	Rotterdam	25.0
Rotterdam	Den Haag	26.0
Den Haag	Hoek van Holland	27.0
Hoek van Holland	Felixstowe	207.0
Felixstowe	Ipswich	22.0
Ipswich	Colchester	32.0
Colchester	London	106.0
London	Doncaster	277.0
Doncaster	Immingham	74.0

If we were in Amsterdam and wanted to visit every other place in our dataset during the same trip, Figure 4-11 demonstrates the shortest continuous route to do so.

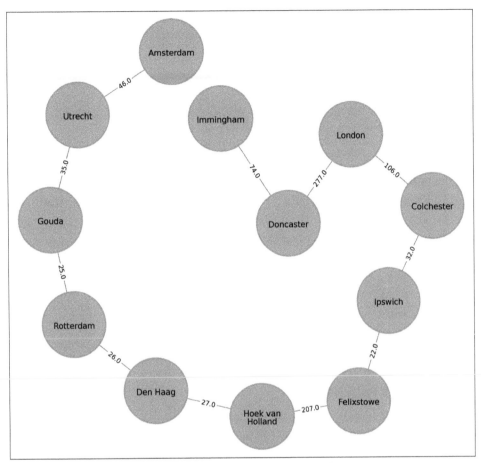

Figure 4-11. A minimum weight spanning tree from Amsterdam

Random Walk

The Random Walk algorithm provides a set of nodes on a random path in a graph. The term was first mentioned by Karl Pearson in 1905 in a letter to *Nature* magazine titled "The Problem of the Random Walk" (*https://go.nature.com/2Fy15em*). Although the concept goes back even further, it's only more recently that random walks have been applied to network science.

A random walk, in general, is sometimes described as being similar to how a drunk person traverses a city. They know what direction or end point they want to reach but may take a very circuitous route to get there.

The algorithm starts at one node and somewhat randomly follows one of the relationships forward or backward to a neighbor node. It then does the same from that node and so on, until it reaches the set path length. (We say somewhat randomly because the number of relationships a node has, and its neighbors have, influences the probability a node will be walked through.)

When Should I Use Random Walk?

Use the Random Walk algorithm as part of other algorithms or data pipelines when you need to generate a mostly random set of connected nodes.

Example use cases include:

- As part of the node2vec and graph2vec algorithms, that create node embeddings. These node embeddings could then be used as the input to a neural network.
- As part of the Walktrap and Infomap community detection. If a random walk returns a small set of nodes repeatedly, then it indicates that node set may have a community structure.
- As part of the training process of machine learning models. This is described further in David Mack's article "Review Prediction with Neo4j and TensorFlow" (*https://bit.ly/2Cx14ph*).

You can read about more use cases in a paper by N. Masuda, M. A. Porter, and R. Lambiotte, "Random Walks and Diffusion on Networks" (*https://bit.ly/2JDvlJ0*).

Random Walk with Neo4j

Neo4j has an implementation of the Random Walk algorithm. It supports two modes for choosing the next relationship to follow at each stage of the algorithm:

random
: Randomly chooses a relationship to follow

node2vec
: Chooses relationship to follow based on computing a probability distribution of the previous neighbors

The Random Walk procedure takes in a config map with the following keys:

start
: The id of the node where our shortest path search begins.

nodeProjection
: Enables the mapping of specific kinds of nodes into the in-memory graph. We can declare one or more node labels.

`relationshipProjection`

Enables the mapping of relationship types into the in-memory graph. We can declare one or more relationship types along with direction and properties.

`walks`

The number of paths returned ``

The following performs a random walk starting from London:

```
MATCH (source:Place {id: "London"})
CALL gds.alpha.randomWalk.stream({
  start: id(source),
  nodeProjection: "*",
  relationshipProjection: {
    all: {
      type: "*",
      properties: "distance",
      orientation:  "UNDIRECTED"
    }
  },
  steps: 5,
  walks: 1
})
YIELD nodeIds
UNWIND gds.util.asNodes(nodeIds) as place
RETURN place.id AS place
```

It returns the following result:

place
London
Doncaster
Immingham
Amsterdam
Utrecht
Amsterdam

At each stage of the random walk the next relationship is chosen randomly. This means that if we rerun the algorithm, even with the same parameters, we likely won't get the same result. It's also possible for a walk to go back on itself, as we can see in Figure 4-12 where we go from Amsterdam to Den Haag and back.

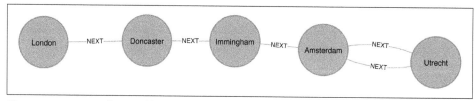

Figure 4-12. A random walk starting from London

Summary

Pathfinding algorithms are useful for understanding the way that our data is connected. In this chapter we started out with the fundamental Breadth and Depth First algorithms, before moving onto Dijkstra and other shortest path algorithms. We also looked at variants of the shortest path algorithms optimized for finding the shortest path from one node to all other nodes or between all pairs of nodes in a graph. We finished with the Random Walk algorithm, which can be used to find arbitrary sets of paths.

Next we'll learn about Centrality algorithms that can be used to find influential nodes in a graph.

Algorithm Resource

There are many algorithm books, but one stands out for its coverage of fundamental concepts and graph algorithms: *The Algorithm Design Manual*, by Steven S. Skiena (Springer). We highly recommend this textbook to those seeking a comprehensive resource on classic algorithms and design techniques, or who simply want to dig deeper into how various algorithms operate.

CHAPTER 5

Centrality Algorithms

Centrality algorithms are used to understand the roles of particular nodes in a graph and their impact on that network. They're useful because they identify the most important nodes and help us understand group dynamics such as credibility, accessibility, the speed at which things spread, and bridges between groups. Although many of these algorithms were invented for social network analysis, they have since found uses in a variety of industries and fields.

We'll cover the following algorithms:

- Degree Centrality as a baseline metric of connectedness
- Closeness Centrality for measuring how central a node is to the group, including two variations for disconnected groups
- Betweenness Centrality for finding control points, including an alternative for approximation
- PageRank for understanding the overall influence, including a popular option for personalization

 Different centrality algorithms can produce significantly different results based on what they were created to measure. When you see suboptimal answers, it's best to check the algorithm you've used is aligned to its intended purpose.

We'll explain how these algorithms work and show examples in Spark and Neo4j. Where an algorithm is unavailable on one platform or where the differences are unimportant, we'll provide just one platform example.

Figure 5-1 shows the differences between the types of questions centrality algorithms can answer, and Table 5-1 is a quick reference for what each algorithm calculates with an example use.

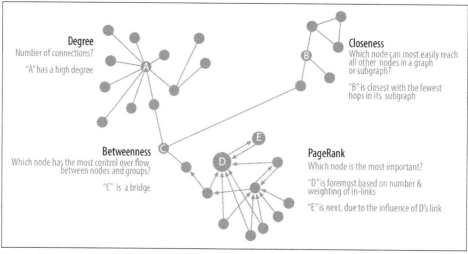

Figure 5-1. Representative centrality algorithms and the types of questions they answer

Table 5-1. Overview of centrality algorithms

Algorithm type	What it does	Example use	Spark example	Neo4j example
Degree Centrality	Measures the number of relationships a node has	Estimating a person's popularity by looking at their in-degree and using their out-degree to estimate gregariousness	Yes	No
Closeness Centrality Variations: Wasserman and Faust, Harmonic Centrality	Calculates which nodes have the shortest paths to all other nodes	Finding the optimal location of new public services for maximum accessibility	Yes	Yes
Betweenness Centrality Variation: Randomized-Approximate Brandes	Measures the number of shortest paths that pass through a node	Improving drug targeting by finding the control genes for specific diseases	No	Yes
PageRank Variation: Personalized PageRank	Estimates a current node's importance from its linked neighbors and their neighbors (popularized by Google)	Finding the most influential features for extraction in machine learning and ranking text for entity relevance in natural language processing.	Yes	Yes

Several of the centrality algorithms calculate shortest paths between every pair of nodes. This works well for small- to medium-sized graphs but for large graphs can be computationally prohibitive. To avoid long runtimes on larger graphs, some algorithms (for example, Betweenness Centrality) have approximating versions.

First, we'll describe the dataset for our examples and walk through importing the data into Apache Spark and Neo4j. Each algorithm is covered in the order listed in Table 5-1. We'll start with a short description of the algorithm and, when warranted, information on how it operates. Variations of algorithms already covered will include less detail. Most sections also include guidance on when to use the related algorithm. We demonstrate example code using a sample dataset at the end of each section.

Let's get started!

Example Graph Data: The Social Graph

Centrality algorithms are relevant to all graphs, but social networks provide a very relatable way to think about dynamic influence and the flow of information. The examples in this chapter are run against a small Twitter-like graph. You can download the nodes and relationships files we'll use to create our graph from the book's GitHub repository (*https://bit.ly/2FPgGVV*).

Table 5-2. social-nodes.csv

id
Alice
Bridget
Charles
Doug
Mark
Michael
David
Amy
James

Table 5-3. social-relationships.csv

src	dst	relationship
Alice	Bridget	FOLLOWS
Alice	Charles	FOLLOWS
Mark	Doug	FOLLOWS
Bridget	Michael	FOLLOWS
Doug	Mark	FOLLOWS
Michael	Alice	FOLLOWS
Alice	Michael	FOLLOWS
Bridget	Alice	FOLLOWS
Michael	Bridget	FOLLOWS

src	dst	relationship
Charles	Doug	FOLLOWS
Bridget	Doug	FOLLOWS
Michael	Doug	FOLLOWS
Alice	Doug	FOLLOWS
Mark	Alice	FOLLOWS
David	Amy	FOLLOWS
James	David	FOLLOWS

Figure 5-2 illustrates the graph that we want to construct.

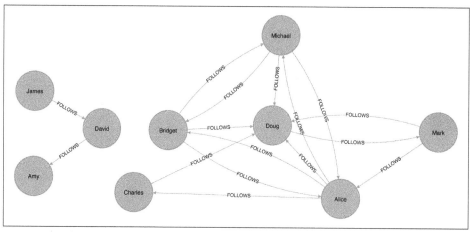

Figure 5-2. The graph model

We have one larger set of users with connections between them and a smaller set with no connections to that larger group.

Let's create graphs in Spark and Neo4j based on the contents of those CSV files.

Importing the Data into Apache Spark

First, we'll import the required packages from Spark and the GraphFrames package:

```
from graphframes import *
from pyspark import SparkContext
```

We can write the following code to create a GraphFrame based on the contents of the CSV files:

```
v = spark.read.csv("data/social-nodes.csv", header=True)
e = spark.read.csv("data/social-relationships.csv", header=True)
g = GraphFrame(v, e)
```

Importing the Data into Neo4j

Next, we'll load the data for Neo4j. We'll start by creating a database that we'll use for the examples in this chapter:

```
:use system;  ❶
CREATE DATABASE chapter5;  ❷
:use chapter5;  ❸
```

❶ Switch to the system database.

❷ Create a new database with the name chapter5. This operation is asynchronous so we may have to wait a couple of seconds before switching to the database.

❸ Switch to the chapter5 database.

The following query imports nodes:

```
WITH 'https://github.com/neo4j-graph-analytics/book/raw/master/data/' AS base
WITH base + 'social-nodes.csv' AS uri
LOAD CSV WITH HEADERS FROM uri AS row
MERGE (:User {id: row.id});
```

And this query imports relationships:

```
WITH 'https://github.com/neo4j-graph-analytics/book/raw/master/data/' AS base
WITH base + 'social-relationships.csv' AS uri
LOAD CSV WITH HEADERS FROM uri AS row
MATCH (source:User {id: row.src})
MATCH (destination:User {id: row.dst})
MERGE (source)-[:FOLLOWS]->(destination);
```

Now that our graphs are loaded, it's on to the algorithms!

Degree Centrality

Degree Centrality is the simplest of the algorithms that we'll cover in this book. It counts the number of incoming and outgoing relationships from a node, and is used to find popular nodes in a graph. Degree Centrality was proposed by Linton C. Free-man in his 1979 paper "Centrality in Social Networks: Conceptual Clarification" (*http://bit.ly/2uAGOih*).

Reach

Understanding the reach of a node is a fair measure of importance. How many other nodes can it touch right now? The *degree* of a node is the number of direct relation-ships it has, calculated for in-degree and out-degree. You can think of this as the immediate reach of node. For example, a person with a high degree in an active social network would have a lot of immediate contacts and be more likely to catch a cold circulating in their network.

The *average degree* of a network is simply the total number of relationships divided by the total number of nodes; it can be heavily skewed by high degree nodes. The *degree distribution* is the probability that a randomly selected node will have a certain number of relationships.

Figure 5-3 illustrates the difference looking at the actual distribution of connections among subreddit topics. If you simply took the average, you'd assume most topics have 10 connections, whereas in fact most topics only have 2 connections.

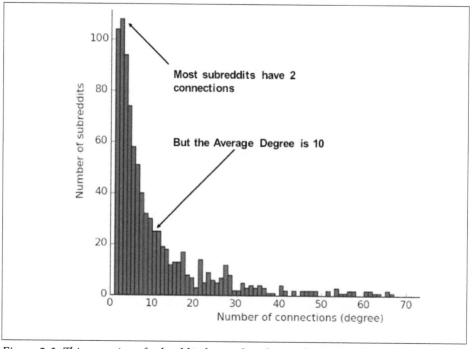

Figure 5-3. This mapping of subreddit degree distribution by Jacob Silterrapa (http://bit.ly/2WlNaOc) provides an example of how the average does not often reflect the actual distribution in networks. CC BY-SA 3.0.

These measures are used to categorize network types such as the scale-free or small-world networks that were discussed in Chapter 2. They also provide a quick measure to help estimate the potential for things to spread or ripple throughout a network.

When Should I Use Degree Centrality?

Use Degree Centrality if you're attempting to analyze influence by looking at the number of incoming and outgoing relationships, or find the "popularity" of individual nodes. It works well when you're concerned with immediate connectedness or near-term probabilities. However, Degree Centrality is also applied to global analysis when you want to evaluate the minimum degree, maximum degree, mean degree, and standard deviation across the entire graph.

Example use cases include:

- Identifying powerful individuals though their relationships, such as connections of people in a social network. For example, in BrandWatch's "Most Influential Men and Women on Twitter 2017" (*https://bit.ly/2WnB2fK*), the top 5 people in each category have over 40 million followers each.
- Separating fraudsters from legitimate users of an online auction site. The weighted centrality of fraudsters tends to be significantly higher due to collusion aimed at artificially increasing prices. Read more in the paper by P. Bangcharoensap et al., "Two Step Graph-Based Semi-Supervised Learning for Online Auction Fraud Detection" (*https://bit.ly/2YlaLAq*).

Degree Centrality with Apache Spark

Now we'll execute the Degree Centrality algorithm with the following code:

```
total_degree = g.degrees
in_degree = g.inDegrees
out_degree = g.outDegrees

(total_degree.join(in_degree, "id", how="left")
 .join(out_degree, "id", how="left")
 .fillna(0)
 .sort("inDegree", ascending=False)
 .show())
```

We first calculate the total, in, and out degrees. Then we join those DataFrames together, using a left join to retain any nodes that don't have incoming or outgoing relationships. If nodes don't have relationships we set that value to 0 using the `fillna` function.

Here's the result of running the code in pyspark:

id	degree	inDegree	outDegree
Doug	6	5	1
Alice	7	3	4
Michael	5	2	3
Bridget	5	2	3
Charles	2	1	1
Mark	3	1	2
David	2	1	1

id	degree	inDegree	outDegree
Amy	1	1	0
James	1	0	1

We can see in Figure 5-4 that Doug is the most popular user in our Twitter graph, with five followers (in-links). All other users in that part of the graph follow him and he only follows one person back. In the real Twitter network, celebrities have high follower counts but tend to follow few people. We could therefore consider Doug a celebrity!

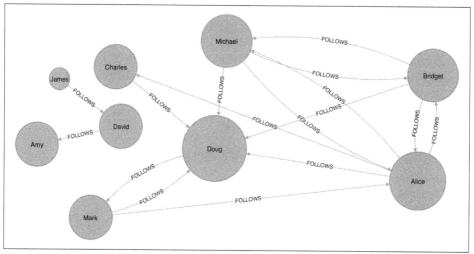

Figure 5-4. Visualization of degree centrality

If we were creating a page showing the most-followed users or wanted to suggest people to follow, we could use this algorithm to identify those people.

 Some data may contain very dense nodes with lots of relationships. These don't add much additional information and can skew some results or add computational complexity. You may want to filter out these dense notes by using a subgraph, or use a projection to summarize the relationships as weights.

Closeness Centrality

Closeness Centrality is a way of detecting nodes that are able to spread information efficiently through a subgraph. The measure of a node's centrality is its average farness (inverse distance) to all other nodes. Nodes with a high closeness score have the shortest distances from all other nodes.

For each node, the Closeness Centrality algorithm calculates the sum of its distances to all other nodes, based on calculating the shortest paths between all pairs of nodes. The resulting sum is then *inverted* to determine the closeness centrality score for that node.

The closeness centrality of a node is calculated using the formula:

$$C(u) = \frac{1}{\sum_{v=1}^{n-1} d(u, v)}$$

where:

- *u* is a node.
- *n* is the number of nodes in the graph.
- *d(u,v)* is the shortest-path distance between another node *v* and *u*.

It is more common to normalize this score so that it represents the average length of the shortest paths rather than their sum. This adjustment allows comparisons of the closeness centrality of nodes of graphs of different sizes.

The formula for normalized closeness centrality is as follows:

$$C_{norm}(u) = \frac{n-1}{\sum_{v=1}^{n-1} d(u, v)}$$

When Should I Use Closeness Centrality?

Apply Closeness Centrality when you need to know which nodes disseminate things the fastest. Using weighted relationships can be especially helpful in evaluating interaction speeds in communication and behavioral analyses.

Example use cases include:

- Uncovering individuals in very favorable positions to control and acquire vital information and resources within an organization. One such study is "Mapping Networks of Terrorist Cells" (*http://bit.ly/2WjFdsM*), by V. E. Krebs.
- As a heuristic for estimating arrival time in telecommunications and package delivery, where content flows through the shortest paths to a predefined target. It is also used to shed light on propagation through all shortest paths simultaneously, such as infections spreading through a local community. Find more details in "Centrality and Network Flow" (*http://bit.ly/2Op5bbH*), by S. P. Borgatti.
- Evaluating the importance of words in a document, based on a graph-based key-phrase extraction process. This process is described by F. Boudin in "A Compari-

son of Centrality Measures for Graph-Based Keyphrase Extraction" (*https://bit.ly/ 2WkDByX*).

 Closeness Centrality works best on connected graphs. When the original formula is applied to an unconnected graph, we end up with an infinite distance between two nodes where there is no path between them. This means that we'll end up with an infinite closeness centrality score when we sum up all the distances from that node. To avoid this issue, a variation on the original formula will be shown after the next example.

Closeness Centrality with Apache Spark

Apache Spark doesn't have a built-in algorithm for Closeness Centrality, but we can write our own using the `aggregateMessages` framework that we introduced in the "Shortest Path (Weighted) with Apache Spark" on page 55 in the previous chapter.

Before we create our function, we'll import some libraries that we'll use:

```
from graphframes.lib import AggregateMessages as AM
from pyspark.sql import functions as F
from pyspark.sql.types import *
from operator import itemgetter
```

We'll also create a few user-defined functions that we'll need later:

```
def collect_paths(paths):
    return F.collect_set(paths)

collect_paths_udf = F.udf(collect_paths, ArrayType(StringType()))

paths_type = ArrayType(
    StructType([StructField("id", StringType()), StructField("distance",

def flatten(ids):
    flat_list = [item for sublist in ids for item in sublist]
    return list(dict(sorted(flat_list, key=itemgetter(0))).items())

flatten_udf = F.udf(flatten, paths_type)

def new_paths(paths, id):
    paths = [{"id": col1, "distance": col2 + 1} for col1,
                        col2 in paths if col1 != id]
    paths.append({"id": id, "distance": 1})
    return paths

new_paths_udf = F.udf(new_paths, paths_type)

def merge_paths(ids, new_ids, id):
```

```
            joined_ids = ids + (new_ids if new_ids else [])
            merged_ids = [(col1, col2) for col1, col2 in joined_ids if col1 != id]
            best_ids = dict(sorted(merged_ids, key=itemgetter(1), reverse=True))
            return [{"id": col1, "distance": col2} for col1, col2 in best_ids.items()]

    merge_paths_udf = F.udf(merge_paths, paths_type)

    def calculate_closeness(ids):
        nodes = len(ids)
        total_distance = sum([col2 for col1, col2 in ids])
        return 0 if total_distance == 0 else nodes * 1.0 / total_distance

    closeness_udf = F.udf(calculate_closeness, DoubleType())
```

And now for the main body that calculates the closeness centrality for each node:

```
    vertices = g.vertices.withColumn("ids", F.array())
    cached_vertices = AM.getCachedDataFrame(vertices)
    g2 = GraphFrame(cached_vertices, g.edges)

    for i in range(0, g2.vertices.count()):
        msg_dst = new_paths_udf(AM.src["ids"], AM.src["id"])
        msg_src = new_paths_udf(AM.dst["ids"], AM.dst["id"])
        agg = g2.aggregateMessages(F.collect_set(AM.msg).alias("agg"),
                                    sendToSrc=msg_src, sendToDst=msg_dst)
        res = agg.withColumn("newIds", flatten_udf("agg")).drop("agg")
        new_vertices = (g2.vertices.join(res, on="id", how="left_outer")
                        .withColumn("mergedIds", merge_paths_udf("ids", "newIds",
                        "id")).drop("ids", "newIds")
                        .withColumnRenamed("mergedIds", "ids"))
        cached_new_vertices = AM.getCachedDataFrame(new_vertices)
        g2 = GraphFrame(cached_new_vertices, g2.edges)

    (g2.vertices
     .withColumn("closeness", closeness_udf("ids"))
     .sort("closeness", ascending=False)
     .show(truncate=False))
```

If we run that we'll see the following output:

id	ids	closeness
Doug	[[Charles, 1], [Mark, 1], [Alice, 1], [Bridget, 1], [Michael, 1]]	1.0
Alice	[[Charles, 1], [Mark, 1], [Bridget, 1], [Doug, 1], [Michael, 1]]	1.0
David	[[James, 1], [Amy, 1]]	1.0
Bridget	[[Charles, 2], [Mark, 2], [Alice, 1], [Doug, 1], [Michael, 1]]	0.7142857142857143
Michael	[[Charles, 2], [Mark, 2], [Alice, 1], [Doug, 1], [Bridget, 1]]	0.7142857142857143
James	[[Amy, 2], [David, 1]]	0.6666666666666666
Amy	[[James, 2], [David, 1]]	0.6666666666666666
Mark	[[Bridget, 2], [Charles, 2], [Michael, 2], [Doug, 1], [Alice, 1]]	0.625

id	ids	closeness
Charles	[[Bridget, 2], [Mark, 2], [Michael, 2], [Doug, 1], [Alice, 1]]	0.625

Alice, Doug, and David are the most closely connected nodes in the graph with a 1.0 score, which means each directly connects to all nodes in their part of the graph. Figure 5-5 illustrates that even though David has only a few connections, within his group of friends that's significant. In other words, this score represents the closeness of each user to others within their subgraph but not the entire graph.

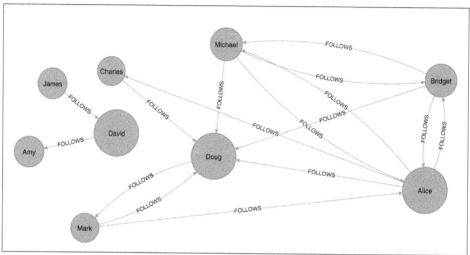

Figure 5-5. Visualization of closeness centrality

Closeness Centrality with Neo4j

Neo4j's implementation of Closeness Centrality uses the following formula:

$$C(u) = \frac{n-1}{\sum_{v=1}^{n-1} d(u, v)}$$

where:

- u is a node.
- n is the number of nodes in the same component (subgraph or group) as u.
- $d(u,v)$ is the shortest-path distance between another node v and u.

The Closeness Centrality algorithm takes in a config map with the following keys:

nodeProjection
> Enables the mapping of specific kinds of nodes into the in-memory graph. We can declare one or more node labels.

relationshipProjection
> Enables the mapping of relationship types into the in-memory graph. We can declare one or more relationship types along with direction and properties.

A call to the following procedure will calculate the closeness centrality using the FOL LOWS relationship type for nodes with the User label:

```
CALL gds.alpha.closeness.stream({
  nodeProjection: "User",
  relationshipProjection: "FOLLOWS"
})
YIELD nodeId, centrality
RETURN gds.util.asNode(nodeId).id, centrality
ORDER BY centrality DESC;
```

Running this procedure gives the following output:

user	centrality
Alice	1.0
Doug	1.0
David	1.0
Bridget	0.7142857142857143
Michael	0.7142857142857143
Amy	0.6666666666666666
James	0.6666666666666666
Charles	0.625
Mark	0.625

We get the same results as with the Spark algorithm, but, as before, the score represents their closeness to others within their subgraph but not the entire graph.

 In the strict interpretation of the Closeness Centrality algorithm, all the nodes in our graph would have a score of ∞ because every node has at least one other node that it's unable to reach. However, it's usually more useful to implement the score per component.

Ideally we'd like to get an indication of closeness across the whole graph, and in the next two sections we'll learn about a few variations of the Closeness Centrality algorithm that do this.

Closeness Centrality Variation: Wasserman and Faust

Stanley Wasserman and Katherine Faust came up with an improved formula for calculating closeness for graphs with multiple subgraphs without connections between those groups. Details on their formula are in their book, *Social Network Analysis: Methods and Applications*. The result of this formula is a ratio of the fraction of nodes in the group that are reachable to the average distance from the reachable nodes.

The formula is as follows:

$$C_{WF}(u) = \frac{n-1}{N-1}\left(\frac{n-1}{\sum_{v=1}^{n-1} d(u,v)}\right)$$

where:

- *u* is a node.
- *N* is the total node count.
- *n* is the number of nodes in the same component as *u*.
- *d(u,v)* is the shortest-path distance between another node *v* and *u*.

We can tell the Closeness Centrality procedure to use this formula by passing the parameter `improved: true`.

The following query executes Closeness Centrality using the Wasserman and Faust formula:

```
CALL gds.alpha.closeness.stream({
  nodeProjection: "User",
  relationshipProjection: "FOLLOWS",
  improved: true
})
YIELD nodeId, centrality
RETURN gds.util.asNode(nodeId).id, centrality
ORDER BY centrality DESC;
```

The procedure gives the following result:

user	centrality
Alice	0.5
Doug	0.5
Bridget	0.35714285714285715
Michael	0.35714285714285715

user	centrality
Charles	0.3125
Mark	0.3125
David	0.125
Amy	0.08333333333333333
James	0.08333333333333333

As Figure 5-6 shows, the results are now more representative of the closeness of nodes to the entire graph. The scores for the members of the smaller subgraph (David, Amy, and James) have been dampened, and they now have the lowest scores of all users. This makes sense as they are the most isolated nodes. This formula is more useful for detecting the importance of a node across the entire graph rather than within its own subgraph.

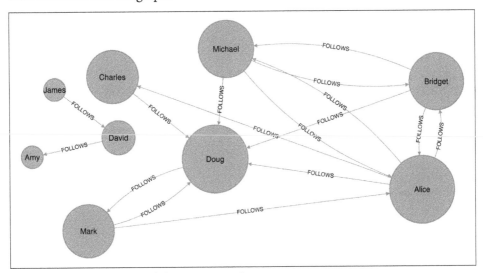

Figure 5-6. Visualization of closeness centrality

In the next section we'll learn about the Harmonic Centrality algorithm, which achieves similar results using another formula to calculate closeness.

Closeness Centrality Variation: Harmonic Centrality

Harmonic Centrality (also known as Valued Centrality) is a variant of Closeness Centrality, invented to solve the original problem with unconnected graphs. In "Harmony in a Small World" (*https://bit.ly/2HSkTef*), M. Marchiori and V. Latora proposed this concept as a practical representation of an average shortest path.

When calculating the closeness score for each node, rather than summing the distances of a node to all other nodes, it sums the inverse of those distances. This means that infinite values become irrelevant.

The raw harmonic centrality for a node is calculated using the following formula:

$$H(u) = \sum_{v=1}^{n-1} \frac{1}{d(u, v)}$$

where:

- u is a node.
- n is the number of nodes in the graph.
- $d(u,v)$ is the shortest-path distance between another node v and u.

As with closeness centrality, we can also calculate a normalized harmonic centrality with the following formula:

$$H_{norm}(u) = \frac{\sum_{v=1}^{n-1} \frac{1}{d(u, v)}}{n - 1}$$

In this formula, ∞ values are handled cleanly.

Harmonic Centrality with Neo4j

The following query executes the Harmonic Centrality algorithm:

```
CALL gds.alpha.closeness.harmonic.stream({
  nodeProjection: "User",
  relationshipProjection: "FOLLOWS"
})
YIELD nodeId, centrality
RETURN gds.util.asNode(nodeId).id, centrality
ORDER BY centrality DESC;
```

Running this procedure gives the following result:

user	centrality
Alice	0.625
Doug	0.625
Bridget	0.5
Michael	0.5
Charles	0.4375
Mark	0.4375
David	0.25
Amy	0.1875
James	0.1875

The results from this algorithm differ from those of the original Closeness Centrality algorithm but are similar to those from the Wasserman and Faust improvement. Either algorithm can be used when working with graphs with more than one connected component.

Betweenness Centrality

Sometimes the most important cog in the system is not the one with the most overt power or the highest status. Sometimes it's the middlemen that connect groups or the brokers who most control over resources or the flow of information. Betweenness Centrality is a way of detecting the amount of influence a node has over the flow of information or resources in a graph. It is typically used to find nodes that serve as a bridge from one part of a graph to another.

The Betweenness Centrality algorithm first calculates the shortest (weighted) path between every pair of nodes in a connected graph. Each node receives a score, based on the number of these shortest paths that pass through the node. The more shortest paths that a node lies on, the higher its score.

Betweenness Centrality was considered one of the "three distinct intuitive conceptions of centrality" when it was introduced by Linton C. Freeman in his 1971 paper, "A Set of Measures of Centrality Based on Betweenness" (*http://moreno.ss.uci.edu/ 23.pdf*).

Bridges and control points

A bridge in a network can be a node or a relationship. In a very simple graph, you can find them by looking for the node or relationship that, if removed, would cause a section of the graph to become disconnected. However, as that's not practical in a typical graph, we use a Betweenness Centrality algorithm. We can also measure the betweenness of a cluster by treating the group as a node.

A node is considered *pivotal* for two other nodes if it lies on *every* shortest path between those nodes, as shown in Figure 5-7.

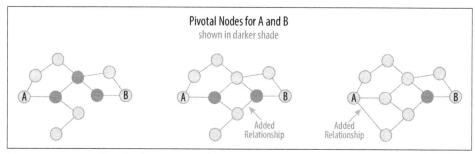

Figure 5-7. Pivotal nodes lie on every shortest path between two nodes. Creating more shortest paths can reduce the number of pivotal nodes for uses such as risk mitigation.

Pivotal nodes play an important role in connecting other nodes—if you remove a pivotal node, the new shortest path for the original node pairs will be longer or more costly. This can be a consideration for evaluating single points of vulnerability.

Calculating betweenness centrality

The betweenness centrality of a node is calculated by adding the results of the following formula for all shortest paths:

$$B(u) = \sum_{s \neq u \neq t} \frac{p(u)}{p}$$

where:

- u is a node.

- p is the total number of shortest paths between nodes s and t.

- $p(u)$ is the number of shortest paths between nodes s and t that pass through node u.

Figure 5-8 illustrates the steps for working out betweenness centrality.

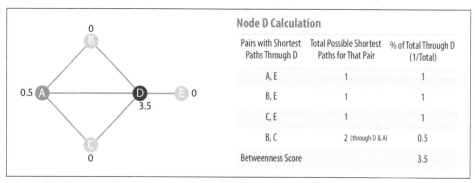

Figure 5-8. Basic concepts for calculating betweenness centrality

---

Here's the procedure:

1. For each node, find the shortest paths that go through it.

   a. B, C, E have no shortest paths and are assigned a value of 0.

2. For each shortest path in step 1, calculate its percentage of the total possible shortest paths for that pair.

3. Add together all the values in step 2 to find a node's betweenness centrality score. The table in Figure 5-8 illustrates steps 2 and 3 for node D.

4. Repeat the process for each node.

When Should I Use Betweenness Centrality?

Betweenness Centrality applies to a wide range of problems in real-world networks. We use it to find bottlenecks, control points, and vulnerabilities.

Example use cases include:

- Identifying influencers in various organizations. Powerful individuals are not necessarily in management positions, but can be found in "brokerage positions" using Betweenness Centrality. Removal of such influencers can seriously destabilize the organization. This might be considered a welcome disruption by law enforcement if the organization is criminal, or could be a disaster if a business loses key staff it underestimated. More details are found in "Brokerage Qualifications in Ringing Operations" (*https://bit.ly/2WKKPg0*), by C. Morselli and J. Roy.

- Uncovering key transfer points in networks such as electrical grids. Counterintuitively, removal of specific bridges can actually *improve* overall robustness by "islanding" disturbances. Research details are included in "Robustness of the European Power Grids Under Intentional Attack" (*https://bit.ly/2Wtqyvp*), by R. Solé, et al.

- Helping microbloggers spread their reach on Twitter, with a recommendation engine for targeting influencers. This approach is described in a paper by S. Wu et al., "Making Recommendations in a Microblog to Improve the Impact of a Focal User" (*https://bit.ly/2Ft58aN*).

Betweenness Centrality makes the assumption that all communication between nodes happens along the shortest path and with the same frequency, which isn't always the case in real life. Therefore, it doesn't give us a perfect view of the most influential nodes in a graph, but rather a good representation. Mark Newman explains this in more detail on p. 186 of *Networks: An Introduction* (*http://bit.ly/2UaM9v0*) (Oxford University Press).

Betweenness Centrality with Neo4j

Spark doesn't have a built-in algorithm for Betweenness Centrality, so we'll demonstrate this algorithm using Neo4j. The Betweenness Centrality algorithm takes in a config map with the following keys:

nodeProjection

> Enables the mapping of specific kinds of nodes into the in-memory graph. We can declare one or more node labels.

relationshipProjection

> Enables the mapping of relationship types into the in-memory graph. We can declare one or more relationship types along with direction and properties.

A call to the following procedure will calculate the betweenness centrality for each of the nodes in our graph:

```
CALL gds.alpha.betweenness.stream({
  nodeProjection: "User",
  relationshipProjection: "FOLLOWS"
})
YIELD nodeId, centrality
RETURN gds.util.asNode(nodeId).id  AS user, centrality
ORDER BY centrality DESC;
```

Running this procedure gives the following result:

user	centrality
Alice	10.0
Doug	7.0
Mark	7.0
David	1.0
Bridget	0.0
Charles	0.0
Michael	0.0
Amy	0.0
James	0.0

As we can see in Figure 5-9, Alice is the main broker in this network, but Mark and Doug aren't far behind. In the smaller subgraph all shortest paths go through David, so he is important for information flow among those nodes.

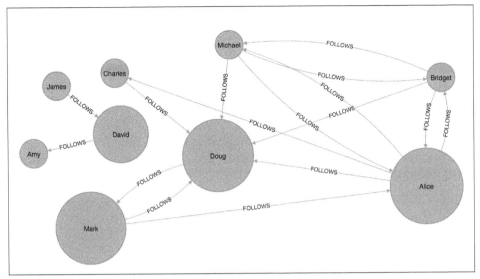

Figure 5-9. Visualization of betweenness centrality

For large graphs, exact centrality computation isn't practical. The fastest known algorithm for exactly computing betweenness of all the nodes has a runtime proportional to the product of the number of nodes and the number of relationships.

We may want to filter down to a subgraph first or use (described in the next section) that works with a subset of nodes.

We can join our two disconnected components together by introducing a new user called Jason, who follows and is followed by people from both groups of users:

```
WITH ["James", "Michael", "Alice", "Doug", "Amy"] AS existingUsers

MATCH (existing:User) WHERE existing.id IN existingUsers
MERGE (newUser:User {id: "Jason"})

MERGE (newUser)<-[:FOLLOWS]-(existing)
MERGE (newUser)-[:FOLLOWS]->(existing)
```

If we rerun the algorithm we'll see this output:

user	centrality
Jason	44.33333333333333
Doug	18.333333333333332
Alice	16.666666666666664
Amy	8.0
James	8.0

user	centrality
Michael	4.0
Mark	2.1666666666666665
David	0.5
Bridget	0.0
Charles	0.0

Jason has the highest score because communication between the two sets of users will pass through him. Jason can be said to act as a *local bridge* between the two sets of users, as illustrated in Figure 5-10.

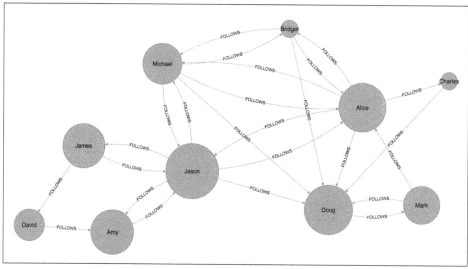

Figure 5-10. Visualization of betweenness centrality with Jason

Before we move on to the next section, let's reset our graph by deleting Jason and his relationships:

```
MATCH (user:User {id: "Jason"})
DETACH DELETE user
```

Betweenness Centrality Variation: Randomized-Approximate Brandes

Recall that calculating the exact betweenness centrality on large graphs can be very expensive. We could therefore choose to use an approximation algorithm that runs much faster but still provides useful (albeit imprecise) information.

The Randomized-Approximate Brandes (RA-Brandes for short) algorithm is the best-known algorithm for calculating an approximate score for betweenness centrality. Rather than calculating the shortest path between every pair of nodes, the RA-Brandes algorithm considers only a subset of nodes. Two common strategies for selecting the subset of nodes are:

Random

Nodes are selected uniformly, at random, with a defined probability of selection. The default probability is: $\frac{log10(N)}{e^2}$. If the probability is 1, the algorithm works the same way as the normal Betweenness Centrality algorithm, where all nodes are loaded.

Degree

Nodes are selected randomly, but those whose degree is lower than the mean are automatically excluded (i.e., only nodes with a lot of relationships have a chance of being visited).

As a further optimization, you could limit the depth used by the Shortest Path algorithm, which will then provide a subset of all the shortest paths.

Approximation of Betweenness Centrality with Neo4j

The following query executes the RA-Brandes algorithm using the random selection method:

```
CALL gds.alpha.betweenness.sampled.stream({
  nodeProjection: "User",
  relationshipProjection: "FOLLOWS",
  strategy: "degree"
})
YIELD nodeId, centrality
RETURN gds.util.asNode(nodeId).id AS user, centrality
ORDER BY centrality DESC;
```

Running this procedure gives the following result:

user	centrality
Alice	9.0
Mark	9.0
Doug	4.5
David	2.25
Bridget	0.0
Charles	0.0
Michael	0.0
Amy	0.0
James	0.0

Our top influencers are similar to before, although Mark now has a higher ranking than Doug.

Due to the random nature of this algorithm, we may see different results each time that we run it. On larger graphs this randomness will have less of an impact than it does on our small sample graph.

PageRank

PageRank is the best known of the centrality algorithms. It measures the transitive (or directional) influence of nodes. All the other centrality algorithms we discuss measure the direct influence of a node, whereas PageRank considers the influence of a node's neighbors, and their neighbors. For example, having a few very powerful friends can make you more influential than having a lot of less powerful friends. PageRank is computed either by iteratively distributing one node's rank over its neighbors or by randomly traversing the graph and counting the frequency with which each node is hit during these walks.

PageRank is named after Google cofounder Larry Page, who created it to rank websites in Google's search results. The basic assumption is that a page with more incoming and more influential incoming links is more likely a credible source. PageRank measures the number and quality of incoming relationships to a node to determine an estimation of how important that node is. Nodes with more sway over a network are presumed to have more incoming relationships from other influential nodes.

Influence

The intuition behind influence is that relationships to more important nodes contribute more to the influence of the node in question than equivalent connections to less important nodes. Measuring influence usually involves scoring nodes, often with weighted relationships, and then updating the scores over many iterations. Sometimes all nodes are scored, and sometimes a random selection is used as a representative distribution.

Keep in mind that centrality measures represent the importance of a node in comparison to other nodes. Centrality is a ranking of the potential impact of nodes, not a measure of actual impact. For example, you might identify the two people with the highest centrality in a network, but perhaps policies or cultural norms are in play that actually shift influence to others. Quantifying actual impact is an active research area to develop additional influence metrics.

The PageRank Formula

PageRank is defined in the original Google paper as follows:

$$PR(u) = (1 - d) + d\left(\frac{PR(T1)}{C(T1)} + \ldots + \frac{PR(Tn)}{C(Tn)}\right)$$

where:

- We assume that a page u has citations from pages $T1$ to Tn.
- d is a damping factor which is set between 0 and 1. It is usually set to 0.85. You can think of this as the probability that a user will continue clicking. This helps minimize rank sink, explained in the next section.
- 1-d is the probability that a node is reached directly without following any relationships.
- $C(Tn)$ is defined as the out-degree of a node T.

Figure 5-11 walks through a small example of how PageRank will continue to update the rank of a node until it converges or meets the set number of iterations.

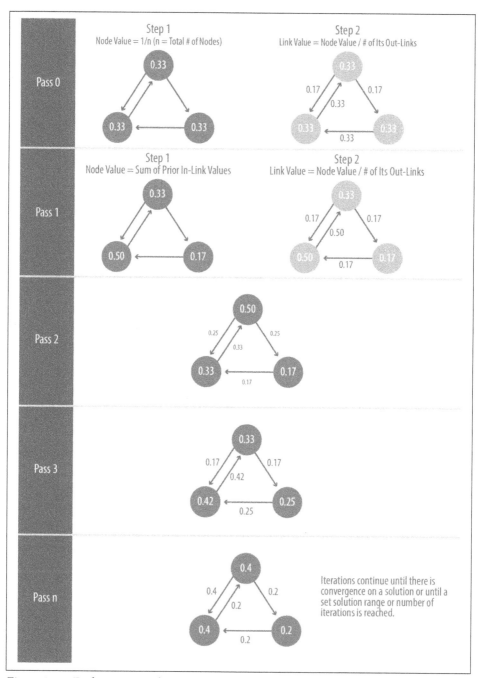

Figure 5-11. Each iteration of PageRank has two calculation steps: one to update node values and one to update link values.

Iteration, Random Surfers, and Rank Sinks

PageRank is an iterative algorithm that runs either until scores converge or until a set number of iterations is reached.

Conceptually, PageRank assumes there is a web surfer visiting pages by following links or by using a random URL. A damping factor _d _ defines the probability that the next click will be through a link. You can think of it as the probability that a surfer will become bored and randomly switch to another page. A PageRank score represents the likelihood that a page is visited through an incoming link and not randomly.

A node, or group of nodes, without outgoing relationships (also called a *dangling node*) can monopolize the PageRank score by refusing to share. This is known as a *rank sink*. You can imagine this as a surfer that gets stuck on a page, or a subset of pages, with no way out. Another difficulty is created by nodes that point only to each other in a group. Circular references cause an increase in their ranks as the surfer bounces back and forth among the nodes. These situations are portrayed in Figure 5-12.

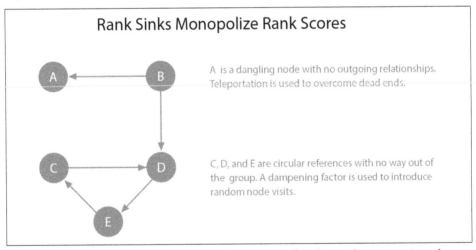

Figure 5-12. Rank sink is caused by a node, or group of nodes, without outgoing relationships.

There are two strategies used to avoid rank sinks. First, when a node is reached that has no outgoing relationships, PageRank assumes outgoing relationships to all nodes. Traversing these invisible links is sometimes called *teleportation*. Second, the damping factor provides another opportunity to avoid sinks by introducing a probability for direct link versus random node visitation. When you set *d* to 0.85, a completely random node is visited 15% of the time.

Although the original formula recommends a damping factor of 0.85, its initial use was on the World Wide Web with a power-law distribution of links (most pages have very few links and a few pages have many). Lowering the damping factor decreases the likelihood of following long relationship paths before taking a random jump. In turn, this increases the contribution of a node's immediate predecessors to its score and rank.

If you see unexpected results from PageRank, it is worth doing some exploratory analysis of the graph to see if any of these problems are the cause. Read Ian Rogers's article, "The Google PageRank Algorithm and How It Works" (*http://bit.ly/2TYSaeQ*) to learn more.

When Should I Use PageRank?

PageRank is now used in many domains outside web indexing. Use this algorithm whenever you're looking for broad influence over a network. For instance, if you're looking to target a gene that has the highest overall impact to a biological function, it may not be the most connected one. It may, in fact, be the gene with the most relationships with other, more significant functions.

Example use cases include:

- Presenting users with recommendations of other accounts that they may wish to follow (Twitter uses Personalized PageRank for this). The algorithm is run over a graph that contains shared interests and common connections. The approach is described in more detail in the paper "WTF: The Who to Follow Service at Twitter" (*https://stanford.io/2ux00wZ*), by P. Gupta et al.

- Predicting traffic flow and human movement in public spaces or streets. The algorithm is run over a graph of road intersections, where the PageRank score reflects the tendency of people to park, or end their journey, on each street. This is described in more detail in "Self-Organized Natural Roads for Predicting Traffic Flow: A Sensitivity Study" (*https://bit.ly/2usHENZ*), a paper by B. Jiang, S. Zhao, and J. Yin.

- As part of anomaly and fraud detection systems in the healthcare and insurance industries. PageRank helps reveal doctors or providers that are behaving in an unusual manner, and the scores are then fed into a machine learning algorithm.

David Gleich describes many more uses for the algorithm in his paper, "PageRank Beyond the Web" (*https://bit.ly/2JCYi80*).

PageRank with Apache Spark

Now we're ready to execute the PageRank algorithm. GraphFrames supports two implementations of PageRank:

- The first implementation runs PageRank for a fixed number of iterations. This can be run by setting the maxIter parameter.
- The second implementation runs PageRank until convergence. This can be run by setting the tol parameter.

PageRank with a fixed number of iterations

Let's see an example of the fixed iterations approach:

```
results = g.pageRank(resetProbability=0.15, maxIter=20)
results.vertices.sort("pagerank", ascending=False).show()
```

 Notice in Spark that the damping factor is more intuitively called the *reset probability*, with the inverse value. In other words, reset Probability=0.15 in this example is equivalent to dampingFac tor:0.85 in Neo4j.

If we run that code in pyspark we'll see this output:

id	pageRank
Doug	2.2865372087512252
Mark	2.1424484186137263
Alice	1.520330830262095
Michael	0.7274429252585624
Bridget	0.7274429252585624
Charles	0.5213852310709753
Amy	0.5097143486157744
David	0.36655842368870073
James	0.1981396884803788

As we might expect, Doug has the highest PageRank because he is followed by all other users in his subgraph. Although Mark only has one follower, that follower is Doug, so Mark is also considered important in this graph. It's not only the number of followers that is important, but also the importance of those followers.

The relationships in the graph on which we ran the PageRank algorithm don't have weights, so each relationship is considered equal. Relationship weights are added by specifying a `weight` column in the relationships DataFrame.

PageRank until convergence

And now let's try the convergence implementation that will run PageRank until it closes in on a solution within the set tolerance:

```
results = g.pageRank(resetProbability=0.15, tol=0.01)
results.vertices.sort("pagerank", ascending=False).show()
```

If we run that code in pyspark we'll see this output:

id	pageRank
Doug	2.2233188859989745
Mark	2.090451188336932
Alice	1.5056291439101062
Michael	0.733738785109624
Bridget	0.733738785109624
Amy	0.559446807245026
Charles	0.5338811076334145
David	0.40232326274180685
James	0.21747203391449021

The PageRank scores for each person are slightly different than with the fixed number of iterations variant, but as we would expect, their order remains the same.

Although convergence on a perfect solution may sound ideal, in some scenarios PageRank cannot mathematically converge. For larger graphs, PageRank execution may be prohibitively long. A tolerance limit helps set an acceptable range for a converged result, but many choose to use (or combine this approach with) the maximum iteration option instead. The maximum iteration setting will generally provide more performance consistency. Regardless of which option you choose, you may need to test several different limits to find what works for your dataset. Larger graphs typcially require more iterations or smaller tolerance than medium-sized graphs for better accuracy.

PageRank with Neo4j

We can also run PageRank in Neo4j. The PageRank algorithm takes in a config map with the following keys:

nodeProjection
> Enables the mapping of specific kinds of nodes into the in-memory graph. We can declare one or more node labels.

relationshipProjection
> Enables the mapping of relationship types into the in-memory graph. We can declare one or more relationship types along with direction and properties.

maxIterations
> The maximum number of iterations of Page Rank to run.

dampingFactor
> The damping factor of the Page Rank calculation.

A call to the following procedure will calculate the PageRank for each of the nodes in our graph:

```
CALL gds.pageRank.stream({
  nodeProjection: "User",
  relationshipProjection: "FOLLOWS",
  maxIterations: 20,
  dampingFactor: 0.85
})
YIELD nodeId, score
RETURN gds.util.asNode(nodeId).id AS page, score
ORDER BY score DESC;
```

Running this procedure gives the following result:

page	score
Doug	1.6704119999999998
Mark	1.5610085
Alice	1.1106700000000003
Bridget	0.535373
Michael	0.535373
Amy	0.385875
Charles	0.3844895
David	0.2775
James	0.15000000000000002

As with the Spark example, Doug is the most influential user, and Mark follows closely after as the only user that Doug follows. We can see the importance of the nodes relative to each other in Figure 5-13.

 PageRank implementations vary, so they can produce different scoring even when the ordering is the same. Neo4j initializes nodes using a value of 1 minus the dampening factor whereas Spark uses a value of 1. In this case, the relative rankings (the goal of PageRank) are identical but the underlying score values used to reach those results are different.

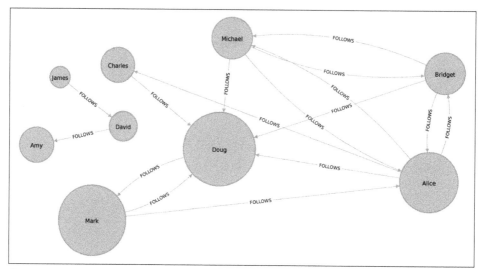

Figure 5-13. Visualization of PageRank

 As with our Spark example, the relationships in the graph on which we ran the PageRank algorithm don't have weights, so each relationship is considered equal. Relationship weights can be considered by including the `weightProperty` property in the config passed to the PageRank procedure. For example, if relationships have a property `weight` containing weights, we would pass the following config to the procedure: `weightProperty: "weight"`.

PageRank Variation: Personalized PageRank

Personalized PageRank (PPR) is a variant of the PageRank algorithm that calculates the importance of nodes in a graph from the perspective of a specific node. For PPR, random jumps refer back to a given set of starting nodes. This biases results toward, or personalizes for, the start node. This bias and localization make PPR useful for highly targeted recommendations.

Personalized PageRank with Apache Spark

We can calculate the personalized PageRank score for a given node by passing in the sourceId parameter. The following code calculates the PPR for Doug:

```
me = "Doug"
results = g.pageRank(resetProbability=0.15, maxIter=20, sourceId=me)
people_to_follow = results.vertices.sort("pagerank", ascending=False)

already_follows = list(g.edges.filter(f"src = '{me}'").toPandas()["dst"])
people_to_exclude = already_follows + [me]

people_to_follow[~people_to_follow.id.isin(people_to_exclude)].show()
```

The results of this query could be used to make recommendations for people who Doug should follow. Notice that we are also making sure that we exclude people who Doug already follows, as well as himself, from our final result.

If we run that code in pyspark we'll see this output:

id	pageRank
Alice	0.1650183746272782
Michael	0.048842467744891996
Bridget	0.048842467744891996
Charles	0.03497796119878669
David	0.0
James	0.0
Amy	0.0

Alice is the best suggestion for somebody that Doug should follow, but we might suggest Michael and Bridget as well.

Summary

Centrality algorithms are an excellent tool for identifying influencers in a network. In this chapter we've learned about the prototypical centrality algorithms: Degree Centrality, Closeness Centrality, Betweenness Centrality, and PageRank. We've also covered several variations to deal with issues such as long runtimes and isolated components, as well as options for alternative uses.

There are many wide-ranging uses for centrality algorithms, and we encourage their exploration for a variety of analyses. You can apply what we've learned to locate optimal touch points for disseminating information, find the hidden brokers that control the flow of resources, and uncover the indirect power players lurking in the shadows.

Next, we'll turn to community detection algorithms that look at groups and partitions.

CHAPTER 6

Community Detection Algorithms

Community formation is common in all types of networks, and identifying them is essential for evaluating group behavior and emergent phenomena. The general principle in finding communities is that its members will have more relationships within the group than with nodes outside their group. Identifying these related sets reveals clusters of nodes, isolated groups, and network structure. This information helps infer similar behavior or preferences of peer groups, estimate resiliency, find nested relationships, and prepare data for other analyses. Community detection algorithms are also commonly used to produce network visualization for general inspection.

We'll provide details on the most representative community detection algorithms:

- Triangle Count and Clustering Coefficient for overall relationship density
- Strongly Connected Components and Weakly Connected Components for finding connected clusters
- Label Propagation for quickly inferring groups based on node labels
- Louvain Modularity for looking at grouping quality and hierarchies

We'll explain how the algorithms work and show examples in Apache Spark and Neo4j. In cases where an algorithm is only available in one platform, we'll provide just one example. We use weighted relationships for these algorithms because they're typically used to capture the significance of different relationships.

Figure 6-1 gives an overview of the differences between the community detection algorithms covered here, and Table 6-1 provides a quick reference as to what each algorithm calculates with example uses.

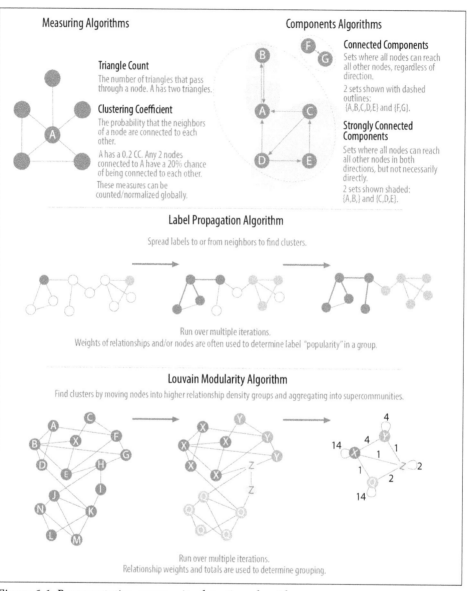

Figure 6-1. Representative community detection algorithms

We use the terms *set*, *partition*, *cluster*, *group*, and *community* interchangeably. These terms are different ways to indicate that similar nodes can be grouped. Community detection algorithms are also called clustering and partitioning algorithms. In each section, we use the terms that are most prominent in the literature for a particular algorithm.

Table 6-1. Overview of community detection algorithms

Algorithm type	What it does	Example use	Spark example	Neo4j example
Triangle Count and Clustering Coefficient	Measures how many nodes form triangles and the degree to which nodes tend to cluster together	Estimating group stability and whether the network might exhibit "small-world" behaviors seen in graphs with tightly knit clusters	Yes	Yes
Strongly Connected Components	Finds groups where each node is reachable from every other node in that same group *following the direction* of relationships	Making product recommendations based on group affiliation or similar items	Yes	Yes
Connected Components	Finds groups where each node is reachable from every other node in that same group, *regardless of the direction* of relationships	Performing fast grouping for other algorithms and identify islands	Yes	Yes
Label Propagation	Infers clusters by spreading labels based on neighborhood majorities	Understanding consensus in social communities or finding dangerous combinations of possible co-prescribed drugs	Yes	Yes
Louvain Modularity	Maximizes the presumed accuracy of groupings by comparing relationship weights and densities to a defined estimate or average	In fraud analysis, evaluating whether a group has just a few discrete bad behaviors or is acting as a fraud ring	No	Yes

First, we'll describe the data for our examples and walk through importing the data into Spark and Neo4j. The algorithms are covered in the order listed in Table 6-1. For each, you'll find a short description and advice on when to use it. Most sections also include guidance on when to use related algorithms. We demonstrate example code using sample data at the end of each algorithm section.

When using community detection algorithms, be conscious of the density of the relationships.

If the graph is very dense, you may end up with all nodes congregating in one or just a few clusters. You can counteract this by filtering by degree, relationship weights, or similarity metrics.

On the other hand, if the graph is too sparse with few connected nodes, you may end up with each node in its own cluster. In this case, try to incorporate additional relationship types that carry more relevant information.

Example Graph Data: The Software Dependency Graph

Dependency graphs are particularly well suited for demonstrating the sometimes subtle differences between community detection algorithms because they tend to be more connected and hierarchical. The examples in this chapter are run against a graph containing dependencies between Python libraries, although dependency graphs are used in various fields, from software to energy grids. This kind of software dependency graph is used by developers to keep track of transitive interdependencies and conflicts in software projects. You can download the nodes and relationships files from the book's GitHub repository (*https://bit.ly/2FPgGVV*).

Table 6-2. sw-nodes.csv

id
six
pandas
numpy
python-dateutil
pytz
pyspark
matplotlib
spacy
py4j
jupyter
jpy-console
nbconvert
ipykernel
jpy-client
jpy-core

Table 6-3. sw-relationships.csv

src	dst	relationship
pandas	numpy	DEPENDS_ON
pandas	pytz	DEPENDS_ON
pandas	python-dateutil	DEPENDS_ON
python-dateutil	six	DEPENDS_ON
pyspark	py4j	DEPENDS_ON
matplotlib	numpy	DEPENDS_ON
matplotlib	python-dateutil	DEPENDS_ON
matplotlib	six	DEPENDS_ON

src	dst	relationship
matplotlib	pytz	DEPENDS_ON
spacy	six	DEPENDS_ON
spacy	numpy	DEPENDS_ON
jupyter	nbconvert	DEPENDS_ON
jupyter	ipykernel	DEPENDS_ON
jupyter	jpy-console	DEPENDS_ON
jpy-console	jpy-client	DEPENDS_ON
jpy-console	ipykernel	DEPENDS_ON
jpy-client	jpy-core	DEPENDS_ON
nbconvert	jpy-core	DEPENDS_ON

Figure 6-2 shows the graph that we want to construct. Looking at this graph, we see that there are three clusters of libraries. We can use visualizations on smaller datasets as a tool to help validate the clusters derived by community detection algorithms.

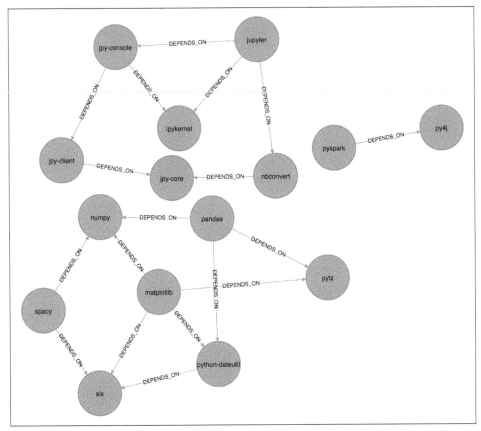

Figure 6-2. The graph model

Let's create graphs in Spark and Neo4j from the example CSV files.

Importing the Data into Apache Spark

We'll first import the packages we need from Apache Spark and the GraphFrames package:

```
from graphframes import *
```

The following function creates a GraphFrame from the example CSV files:

```
def create_software_graph():
    nodes = spark.read.csv("data/sw-nodes.csv", header=True)
    relationships = spark.read.csv("data/sw-relationships.csv", header=True)
    return GraphFrame(nodes, relationships)
```

Now let's call that function:

```
g = create_software_graph()
```

Importing the Data into Neo4j

Next we'll do the same for Neo4j. We'll start by creating a database that we'll use for the examples in this chapter:

```
:use system;  ❶
CREATE DATABASE chapter6;  ❷
:use chapter6;  ❸
```

❶ Switch to the system database.

❷ Create a new database with the name chapter6. This operation is asynchronous so we may have to wait a couple of seconds before switching to the database.

❸ Switch to the chapter6 database.

The following query imports the nodes:

```
WITH 'https://github.com/neo4j-graph-analytics/book/raw/master/data/' AS base
WITH base + 'sw-nodes.csv' AS uri
LOAD CSV WITH HEADERS FROM uri AS row
MERGE (:Library {id: row.id});
```

And this imports the relationships:

```
WITH 'https://github.com/neo4j-graph-analytics/book/raw/master/data/' AS base
WITH base + 'sw-relationships.csv' AS uri
LOAD CSV WITH HEADERS FROM uri AS row
MATCH (source:Library {id: row.src})
MATCH (destination:Library {id: row.dst})
MERGE (source)-[:DEPENDS_ON]->(destination);
```

Now that we've got our graphs loaded it's on to the algorithms!

Triangle Count and Clustering Coefficient

The Triangle Count and Clustering Coefficient algorithms are presented together because they are so often used together. Triangle Count determines the number of triangles passing through each node in the graph. A triangle is a set of three nodes, where each node has a relationship to all other nodes. Triangle Count can also be run globally for evaluating our overall dataset.

> Networks with a high number of triangles are more likely to exhibit small-world structures and behaviors.

The goal of the Clustering Coefficient algorithm is to measure how tightly a group is clustered compared to how tightly it could be clustered. The algorithm uses Triangle Count in its calculations, which provides a ratio of existing triangles to possible relationships. A maximum value of 1 indicates a clique where every node is connected to every other node.

There are two types of clustering coefficients: local clustering and global clustering.

Local Clustering Coefficient

The local clustering coefficient of a node is the likelihood that its neighbors are also connected. The computation of this score involves triangle counting.

The clustering coefficient of a node can be found by multiplying the number of triangles passing through the node by two and then diving that by the maximum number of relationships in the group, which is always the degree of that node, minus one. Examples of different triangles and clustering coefficients for a node with five relationships are portrayed in Figure 6-3.

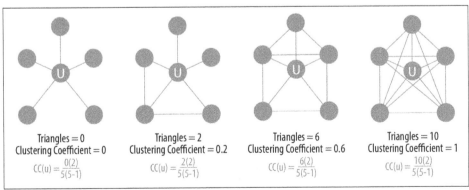

Figure 6-3. *Triangle counts and clustering coefficients for node u*

Note in Figure 6-3, we use a node with five relationships which makes it appear that the clustering coefficient will always equate to 10% of the number of triangles. We can see this is not the case when we alter the number of relationships. If we change the second example to have four relationships (and the same two triangles) then the coefficient is 0.33.

The clustering coefficient for a node uses the formula:

$$CC(u) = \frac{2R_u}{k_u(k_u - 1)}$$

where:

- u is a node.
- $R(u)$ is the number of relationships through the neighbors of u (this can be obtained by using the number of triangles passing through u).
- $k(u)$ is the degree of u.

Global Clustering Coefficient

The global clustering coefficient is the normalized sum of the local clustering coefficients.

Clustering coefficients give us an effective means to find obvious groups like cliques, where every node has a relationship with all other nodes, but we can also specify thresholds to set levels (say, where nodes are 40% connected).

When Should I Use Triangle Count and Clustering Coefficient?

Use Triangle Count when you need to determine the stability of a group or as part of calculating other network measures such as the clustering coefficient. Triangle counting is popular in social network analysis, where it is used to detect communities.

Clustering Coefficient can provide the probability that randomly chosen nodes will be connected. You can also use it to quickly evaluate the cohesiveness of a specific group or your overall network. Together these algorithms are used to estimate resiliency and look for network structures.

Example use cases include:

- Identifying features for classifying a given website as spam content. This is described in "Efficient Semi-Streaming Algorithms for Local Triangle Counting in Massive Graphs" (*http://bit.ly/2ut0Lao*), a paper by L. Becchetti et al.
- Investigating the community structure of Facebook's social graph, where researchers found dense neighborhoods of users in an otherwise sparse global

graph. Find this study in the paper "The Anatomy of the Facebook Social Graph" (*https://bit.ly/2TXWsTC*), by J. Ugander et al.

- Exploring the thematic structure of the web and detecting communities of pages with common topics based on the reciprocal links between them. For more information, see "Curvature of Co-Links Uncovers Hidden Thematic Layers in the World Wide Web" (*http://bit.ly/2YkCrFo*), by J.-P. Eckmann and E. Moses.

Triangle Count with Apache Spark

Now we're ready to execute the Triangle Count algorithm. We can use the following code to do this:

```
result = g.triangleCount()
(result.sort("count", ascending=False)
 .filter('count > 0')
 .show())
```

If we run that code in pyspark we'll see this output:

count	id
1	jupyter
1	python-dateutil
1	six
1	ipykernel
1	matplotlib
1	jpy-console

A triangle in this graph would indicate that two of a node's neighbors are also neighbors. Six of our libraries participate in such triangles.

What if we want to know which nodes are in those triangles? That's where a *triangle stream* comes in. For this, we need Neo4j.

Triangles with Neo4j

Getting a stream of the triangles isn't available using Spark, but we can return it using Neo4j. The triangles algorithm takes in a config map with the following keys:

nodeProjection
: Enables the mapping of specific kinds of nodes into the in-memory graph. We can declare one or more node labels.

relationshipProjection
: Enables the mapping of relationship types into the in-memory graph. We can declare one or more relationship types along with direction and properties.

The Triangles and Triangle Count procedures require relationship projections to be undirected, which we can do by setting `orienta tion: "UNDIRECTED"`.

A call to the following procedure will calculate the triangles in the graph:

```
CALL gds.alpha.triangles({
  nodeProjection: "Library",
  relationshipProjection: {
    DEPENDS_ON: {
      type: "DEPENDS_ON",
      orientation: "UNDIRECTED"
    }
  }
})
YIELD nodeA, nodeB, nodeC
RETURN gds.util.asNode(nodeA).id AS nodeA,
       gds.util.asNode(nodeB).id AS nodeB,
       gds.util.asNode(nodeC).id AS nodeC;
```

Running this procedure gives the following result:

nodeA	nodeB	nodeC
matplotlib	six	python-dateutil
jupyter	jpy-console	ipykernel

We see the same six libraries as we did before, but now we know how they're connected. matplotlib, six, and python-dateutil form one triangle. jupyter, jpy-console, and ipykernel form the other. We can see these triangles in Figure 6-4.

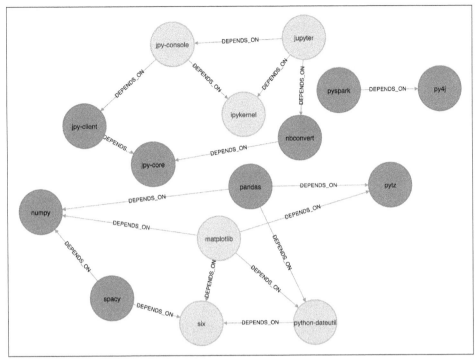

Figure 6-4. Triangles in the software dependency graph

Local Clustering Coefficient with Neo4j

We can also work out the local clustering coefficient using the triangle count algorithm. The triangle count algorithm takes in a config map with the following keys:

nodeProjection
> Enables the mapping of specific kinds of nodes into the in-memory graph. We can declare one or more node labels.

relationshipProjection
> Enables the mapping of relationship types into the in-memory graph. We can declare one or more relationship types along with direction and properties.

The following query calculates the local clustering coefficient for each node:

```
CALL gds.localClusteringCoefficient.stream({
  nodeProjection: "Library",
  relationshipProjection: {
    DEPENDS_ON: {
      type: "DEPENDS_ON",
      orientation: "UNDIRECTED"
    }
  }
}
```

```
})
YIELD nodeId, localClusteringCoefficient
WHERE localClusteringCoefficient > 0
RETURN gds.util.asNode(nodeId).id AS library, localClusteringCoefficient
ORDER BY localClusteringCoefficient DESC;
```

Running this procedure gives the following result:

library	localClusteringCoefficient
ipykernel	1.0
jupyter	0.3333333333333333
jpy-console	0.3333333333333333
six	0.3333333333333333
python-dateutil	0.3333333333333333
matplotlib	0.16666666666666666

ipykernel has a score of 1, which means that all ipykernel's neighbors are neighbors of each other. We can clearly see that in Figure 6-4. This tells us that the community directly around ipykernel is very cohesive.

We've filtered out nodes with a coefficient score of 0 in this code sample, but nodes with low coefficients may also be interesting. A low score can be an indicator that a node is a *structural hole* (*http://stanford.io/2UTYVex*)—a node that is well connected to nodes in different communities that aren't otherwise connected to each other. This is a method for finding *potential* bridges that we discussed in Chapter 5.

Strongly Connected Components

The Strongly Connected Components (SCC) algorithm is one of the earliest graph algorithms. SCC finds sets of connected nodes in a directed graph where each node is reachable in both directions from any other node in the same set. Its runtime operations scale well, proportional to the number of nodes. In Figure 6-5 you can see that the nodes in an SCC group don't need to be immediate neighbors, but there must be directional paths between all nodes in the set.

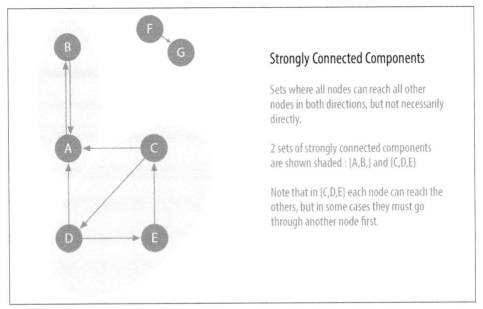

Figure 6-5. Strongly connected components

Decomposing a directed graph into its strongly connected compo-
nents is a classic application of the Depth First Search algorithm.
Neo4j uses DFS under the hood as part of its implementation of the
SCC algorithm.

When Should I Use Strongly Connected Components?

Use Strongly Connected Components as an early step in graph analysis to see how a
graph is structured or to identify tight clusters that may warrant independent investigation. A component that is strongly connected can be used to profile similar behavior or inclinations in a group for applications such as recommendation engines.

Many community detection algorithms like SCC are used to find and collapse clusters
into single nodes for further intercluster analysis. You can also use SCC to visualize
cycles for analyses like finding processes that might deadlock because each subpro-
cess is waiting for another member to take action.

Example use cases include:

- Finding the set of firms in which every member directly and/or indirectly owns
  shares in every other member, as in "The Network of Global Corporate Control"
  (*http://bit.ly/2UU4EAP*), an analysis of powerful transnational corporations by S.
  Vitali, J. B. Glattfelder, and S. Battiston.

- Computing the connectivity of different network configurations when measuring routing performance in multihop wireless networks. Read more in "Routing Performance in the Presence of Unidirectional Links in Multihop Wireless Networks" (*https://bit.ly/2uAJs7H*), by M. K. Marina and S. R. Das.

- Acting as the first step in many graph algorithms that work only on strongly connected graphs. In social networks we find many strongly connected groups. In these sets people often have similar preferences, and the SCC algorithm is used to find such groups and suggest pages to like or products to purchase to the people in the group who have not yet done so.

Some algorithms have strategies for escaping infinite loops, but if we're writing our own algorithms or finding nonterminating processes, we can use SCC to check for cycles.

Strongly Connected Components with Apache Spark

Starting with Apache Spark, we'll first import the packages we need from Spark and the GraphFrames package:

```
from graphframes import *
from pyspark.sql import functions as F
```

Now we're ready to execute the Strongly Connected Components algorithm. We'll use it to work out whether there are any circular dependencies in our graph.

Two nodes can only be in the same strongly connected component if there are paths between them in both directions.

We write the following code to do this:

```
result = g.stronglyConnectedComponents(maxIter=10)
(result.sort("component")
 .groupby("component")
 .agg(F.collect_list("id").alias("libraries"))
 .show(truncate=False))
```

If we run that code in pyspark we'll see this output:

component	libraries
180388626432	[jpy-core]
223338299392	[spacy]
498216206336	[numpy]
523986010112	[six]
549755813888	[pandas]
558345748480	[nbconvert]
661424963584	[ipykernel]
721554505728	[jupyter]
764504178688	[jpy-client]
833223655424	[pytz]
910533066752	[python-dateutil]
936302870528	[pyspark]
944892805120	[matplotlib]
1099511627776	[jpy-console]
1279900254208	[py4j]

You might notice that every library node is assigned to a unique component. This is the partition or subgroup it belongs to, and as we (hopefully!) expected, every node is in its own partition. This means our software project has no circular dependencies amongst these libraries.

Strongly Connected Components with Neo4j

Let's run the same algorithm using Neo4j. The Strongly Connected Components algorithm takes in a config map with the following keys:

nodeProjection

Enables the mapping of specific kinds of nodes into the in-memory graph. We can declare one or more node labels.

relationshipProjection

Enables the mapping of relationship types into the in-memory graph. We can declare one or more relationship types along with direction and properties.

We can execute the following query to run the algorithm:

```
CALL gds.alpha.scc.stream({
  nodeProjection: "Library",
  relationshipProjection: "DEPENDS_ON"
})
YIELD nodeId, partition
```

```
    RETURN partition, collect(gds.util.asNode(nodeId).id) AS libraries
    ORDER BY size(libraries) DESC;
```

The parameters passed to this algorithm are:

Library
    The node label to load from the graph

DEPENDS_ON
    The relationship type to load from the graph

This is the output we'll see when we run the query:

partition	libraries
8	[ipykernel]
11	[six]
2	[matplotlib]
5	[jupyter]
14	[python-dateutil]
13	[numpy]
4	[py4j]
7	[nbconvert]
1	[pyspark]
10	[jpy-core]
9	[jpy-client]
3	[spacy]
12	[pandas]
6	[jpy-console]
0	[pytz]

As with the Spark example, every node is in its own partition.

So far the algorithm has only revealed that our Python libraries are very well behaved, but let's create a circular dependency in the graph to make things more interesting. This should mean that we'll end up with some nodes in the same partition.

The following query adds an extra library that creates a circular dependency between py4j and pyspark:

```
    MATCH (py4j:Library {id: "py4j"})
    MATCH (pyspark:Library {id: "pyspark"})
    MERGE (extra:Library {id: "extra"})
    MERGE (py4j)-[:DEPENDS_ON]->(extra)
    MERGE (extra)-[:DEPENDS_ON]->(pyspark);
```

We can clearly see the circular dependency that got created in Figure 6-6.

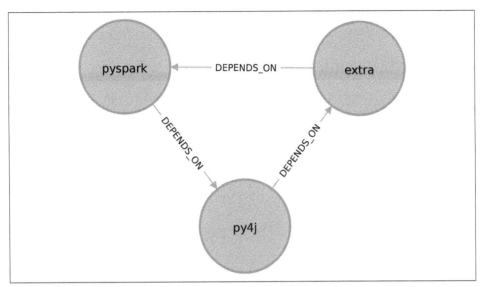

Figure 6-6. A circular dependency between pyspark, py4j, and extra

Now if we run the SCC algorithm again we'll see a slightly different result:

partition	libraries
1	[pyspark, py4j, extra]
8	[ipykernel]
11	[six]
2	[matplotlib]
5	[jupyter]
14	[numpy]
13	[pandas]
7	[nbconvert]
10	[jpy-core]
9	[jpy-client]
3	[spacy]
15	[python-dateutil]
6	[jpy-console]
0	[pytz]

pyspark, py4j, and extra are all part of the same partition, and SCCs helped us find the circular dependency!

Before we move on to the next algorithm we'll delete the extra library and its relationships from the graph:

```
MATCH (extra:Library {id: "extra"})
DETACH DELETE extra;
```

Connected Components

The Connected Components algorithm (sometimes called Union Find or Weakly Connected Components) finds sets of connected nodes in an undirected graph where each node is reachable from any other node in the same set. It differs from the SCC algorithm because it only needs a path to exist between pairs of nodes in one direction, whereas SCC needs a path to exist in both directions. Bernard A. Galler and Michael J. Fischer first described this algorithm in their 1964 paper, "An Improved Equivalence Algorithm" (*https://bit.ly/2WsPNxT*).

When Should I Use Connected Components?

As with SCC, Connected Components is often used early in an analysis to understand a graph's structure. Because it scales efficiently, consider this algorithm for graphs requiring frequent updates. It can quickly show new nodes in common between groups, which is useful for analysis such as fraud detection.

Make it a habit to run Connected Components to test whether a graph is connected as a preparatory step for general graph analysis. Performing this quick test can avoid accidentally running algorithms on only one disconnected component of a graph and getting incorrect results.

Example use cases include:

- Keeping track of clusters of database records, as part of the deduplication process. Deduplication is an important task in master data management applications; the approach is described in more detail in "An Efficient Domain-Independent Algorithm for Detecting Approximately Duplicate Database Records" (*http://bit.ly/2CCNpgy*), by A. Monge and C. Elkan.
- Analyzing citation networks. One study uses Connected Components to work out how well connected a network is, and then to see whether the connectivity remains if "hub" or "authority" nodes are moved from the graph. This use case is explained further in "Characterizing and Mining Citation Graph of Computer Science Literature" (*https://bit.ly/2U8cfi9*), a paper by Y. An, J. C. M. Janssen, and E. E. Milios.

Connected Components with Apache Spark

Starting with Apache Spark, we'll first import the packages we need from Spark and the GraphFrames package:

```
from pyspark.sql import functions as F
```

 Two nodes can be in the same connected component if there is a path between them in either direction.

Now we're ready to execute the Connected Components algorithm. We write the following code to do this:

```
result = g.connectedComponents()
(result.sort("component")
 .groupby("component")
 .agg(F.collect_list("id").alias("libraries"))
 .show(truncate=False))
```

If we run that code in pyspark we'll see this output:

component	libraries
180388626432	[jpy-core, nbconvert, ipykernel, jupyter, jpy-client, jpy-console]
223338299392	[spacy, numpy, six, pandas, pytz, python-dateutil, matplotlib]
936302870528	[pyspark, py4j]

The results show three clusters of nodes, which can also be seen in Figure 6-7. In this example it's very easy to see that there are three components just by visual inspection. This algorithm shows its value more on larger graphs, where visual inspection isn't possible or is very time-consuming.

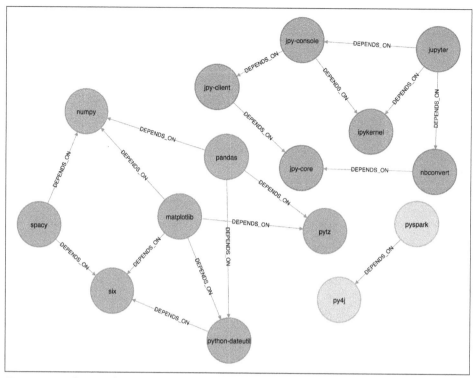

Figure 6-7. Clusters found by the Connected Components algorithm

Connected Components with Neo4j

We can also execute this algorithm in Neo4j. The Connected Components algorithm takes in a config map with the following keys:

nodeProjection
> Enables the mapping of specific kinds of nodes into the in-memory graph. We can declare one or more node labels.

relationshipProjection
> Enables the mapping of relationship types into the in-memory graph. We can declare one or more relationship types along with direction and properties.

We can execute this algorithm by running the following query:

```
CALL gds.wcc.stream({
  nodeProjection: "Library",
  relationshipProjection: "DEPENDS_ON"
})
YIELD nodeId, componentId
RETURN componentId, collect(gds.util.asNode(nodeId).id) AS libraries
ORDER BY size(libraries) DESC;
```

Here's the output:

componentId	libraries
2	[pytz, matplotlib, spacy, six, pandas, numpy, python-dateutil]
5	[jupyter, jpy-console, nbconvert, ipykernel, jpy-client, jpy-core]
1	[pyspark, py4j]

As expected, we get exactly the same results as we did with Spark.

Both of the community detection algorithms that we've covered so far are deterministic: they return the same results each time we run them. Our next two algorithms are examples of nondeterministic algorithms, where we may see different results if we run them multiple times, even on the same data.

Label Propagation

The Label Propagation algorithm (LPA) is a fast algorithm for finding communities in a graph. In LPA, nodes select their group based on their direct neighbors. This process is well suited to networks where groupings are less clear and weights can be used to help a node determine which community to place itself within. It also lends itself well to semisupervised learning because you can seed the process with preassigned, indicative node labels.

The intuition behind this algorithm is that a single label can quickly become dominant in a densely connected group of nodes, but it will have trouble crossing a sparsely connected region. Labels get trapped inside a densely connected group of nodes, and nodes that end up with the same label when the algorithm finishes are considered part of the same community. The algorithm resolves overlaps, where nodes are potentially part of multiple clusters, by assigning membership to the label neighborhood with the highest combined relationship and node weight. LPA is a relatively new algorithm proposed in 2007 by U. N. Raghavan, R. Albert, and S. Kumara, in a paper titled "Near Linear Time Algorithm to Detect Community Structures in Large-Scale Networks" (*https://bit.ly/2Frb1Fu*).

Figure 6-8 depicts two variations of Label Propagation, a simple push method and the more typical pull method that relies on relationship weights. The pull method lends itself well to parallelization.

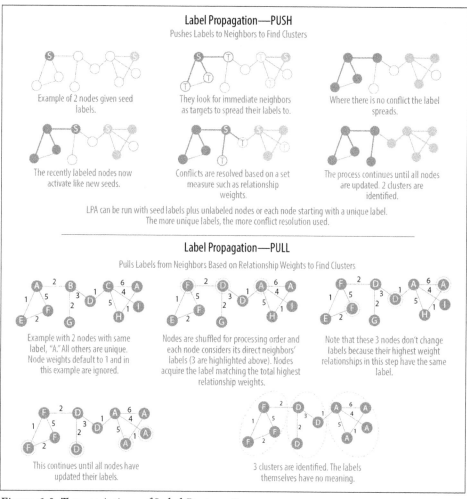

Figure 6-8. Two variations of Label Propagation

The steps often used for the Label Propagation pull method are:

1. Every node is initialized with a unique label (an identifier), and, optionally preliminary "seed" labels can be used.

2. These labels propagate through the network.

3. At every propagation iteration, each node updates its label to match the one with the maximum weight, which is calculated based on the weights of neighbor nodes *and* their relationships. Ties are broken uniformly and randomly.

4. LPA reaches convergence when each node has the majority label of its neighbors.

As labels propagate, densely connected groups of nodes quickly reach a consensus on a unique label. At the end of the propagation, only a few labels will remain, and nodes that have the same label belong to the same community.

Semi-Supervised Learning and Seed Labels

In contrast to other algorithms, Label Propagation can return different community structures when run multiple times on the same graph. The order in which LPA evaluates nodes can have an influence on the final communities it returns.

The range of solutions is narrowed when some nodes are given preliminary labels (i.e., seed labels), while others are unlabeled. Unlabeled nodes are more likely to adopt the preliminary labels.

This use of Label Propagation can be considered a *semi-supervised learning* method to find communities. Semi-supervised learning is a class of machine learning tasks and techniques that operate on a small amount of labeled data, along with a larger amount of unlabeled data. We can also run the algorithm repeatedly on graphs as they evolve.

Finally, LPA sometimes doesn't converge on a single solution. In this situation, our community results will continually flip between a few remarkably similar communities and the algorithm would never complete. Seed labels help guide it toward a solution. Spark and Neo4j use a set maximum number of iterations to avoid never-ending execution. You should test the iteration setting for your data to balance accuracy and execution time.

When Should I Use Label Propagation?

Use Label Propagation in large-scale networks for initial community detection, especially when weights are available. This algorithm can be parallelized and is therefore extremely fast at graph partitioning.

Example use cases include:

- Assigning polarity of tweets as a part of semantic analysis. In this scenario, positive and negative seed labels from a classifier are used in combination with the Twitter follower graph. For more information, see "Twitter Polarity Classification with Label Propagation over Lexical Links and the Follower Graph" (*https://bit.ly/2FBq2pv*), by M. Speriosu et al.

- Finding potentially dangerous combinations of possible co-prescribed drugs, based on the chemical similarity and side effect profiles. See "Label Propagation Prediction of Drug–Drug Interactions Based on Clinical Side Effects" (*https://www.nature.com/articles/srep12339*), a paper by P. Zhang et al.

- Inferring dialogue features and user intention for a machine learning model. For more information, see "Feature Inference Based on Label Propagation on Wikidata Graph for DST" (*https://bit.ly/2FtGpTK*), a paper by Y. Murase et al.

Label Propagation with Apache Spark

Starting with Apache Spark, we'll first import the packages we need from Spark and the GraphFrames package:

```
from pyspark.sql import functions as F
```

Now we're ready to execute the Label Propagation algorithm. We write the following code to do this:

```
result = g.labelPropagation(maxIter=10)
(result
.sort("label")
.groupby("label")
.agg(F.collect_list("id"))
.show(truncate=False))
```

If we run that code in pyspark, we'll see this output:

label	collect_list(id)
180388626432	[jpy-core, jpy-console, jupyter]
223338299392	[matplotlib, spacy]
498216206336	[python-dateutil, numpy, six, pytz]
549755813888	[pandas]
558345748480	[nbconvert, ipykernel, jpy-client]
936302870528	[pyspark]
1279900254208	[py4j]

Compared to Connected Components, we have more clusters of libraries in this example. LPA is less strict than Connected Components with respect to how it determines clusters. Two neighbors (directly connected nodes) may be found to be in different clusters using Label Propagation. However, using Connected Components a node would always be in the same cluster as its neighbors because that algorithm bases grouping strictly on relationships.

In our example, the most obvious difference is that the Jupyter libraries have been split into two communities—one containing the core parts of the library and the other the client-facing tools.

Label Propagation with Neo4j

Now let's try the same algorithm with Neo4j. LPA takes in a config map with the following keys:

nodeProjection
> Enables the mapping of specific kinds of nodes into the in-memory graph. We can declare one or more node labels.

relationshipProjection
> Enables the mapping of relationship types into the in-memory graph. We can declare one or more relationship types along with direction and properties.

maxIterations
> The maximum number of iterations of Label Propagation to run.

We can execute LPA by running the following query:

```
CALL gds.labelPropagation.stream({
  nodeProjection: "Library",
  relationshipProjection: "DEPENDS_ON",
  maxIterations: 10
})
YIELD nodeId, communityId
RETURN communityId AS label,
       collect(gds.util.asNode(nodeId).id) AS libraries
ORDER BY size(libraries) DESC;
```

If we run this procedure, these are the results we'd see:

label	libraries
32	["six", "pandas", "python-dateutil", "matplotlib", "spacy"]
0	["ipykernel", "jupyter", "jpy-console"]
2	["jpy-client", "jpy-core", "nbconvert"]
908	["pyspark", "py4j"]
34	["numpy"]
36	["pytz"]

A subset of the results can be seen visually in Figure 6-9 and are similar to those we got with Apache Spark.

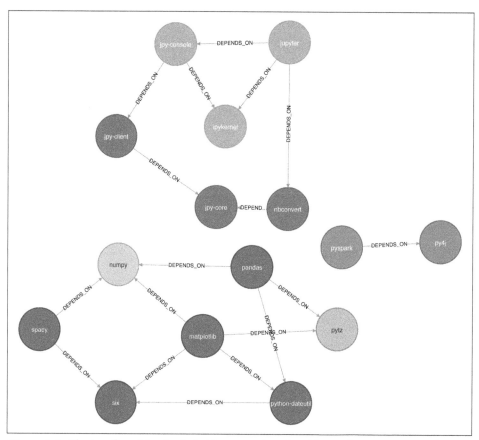

Figure 6-9. Clusters found by the Label Propagation algorithm, when respecting relationship direction

We can also run the algorithm assuming that the graph is undirected, which means that nodes will try to adopt labels from the libraries they depend on as well as ones that depend on them.

To do this, we pass the `DIRECTION:BOTH` parameter to the algorithm:

```
CALL gds.labelPropagation.stream({
  nodeProjection: "Library",
  relationshipProjection: {
    DEPENDS_ON: {
      type: "DEPENDS_ON",
      orientation: "UNDIRECTED"
    }
  },
  maxIterations: 10
})
YIELD nodeId, communityId
```

```
RETURN communityId AS label,
       collect(gds.util.asNode(nodeId).id) AS libraries
ORDER BY size(libraries) DESC;
```

If we run that, we'll get the following output:

label	libraries
11	[pytz, matplotlib, spacy, six, pandas, numpy, python-dateutil]
10	[nbconvert, jpy-client, jpy-core]
6	[jupyter, jpy-console, ipykernel]
4	[pyspark, py4j]

The number of clusters has reduced from six to four, and all the nodes in the matplotlib part of the graph are now grouped together. This can be seen more clearly in Figure 6-10.

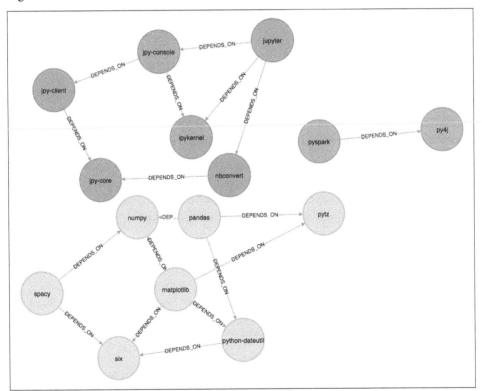

Figure 6-10. Clusters found by the Label Propagation algorithm, when ignoring relationship direction

Although the results of running Label Propagation on this data are similar for undirected and directed calculation, on complicated graphs you will see more significant differences. This is because ignoring direction causes nodes to try and adopt more labels, regardless of the relationship source.

Louvain Modularity

The Louvain Modularity algorithm finds clusters by comparing community density as it assigns nodes to different groups. You can think of this as a "what if" analysis to try various groupings with the goal of reaching a global optimum.

Proposed in 2008, the Louvain algorithm (*https://arxiv.org/pdf/0803.0476.pdf*) is one of the fastest modularity-based algorithms. As well as detecting communities, it also reveals a hierarchy of communities at different scales. This is useful for understanding the structure of a network at different levels of granularity.

Louvain quantifies how well a node is assigned to a group by looking at the density of connections within a cluster in comparison to an average or random sample. This measure of community assignment is called *modularity*.

Quality-based grouping via modularity

Modularity is a technique for uncovering communities by partitioning a graph into more coarse-grained modules (or clusters) and then measuring the strength of the groupings. As opposed to just looking at the concentration of connections within a cluster, this method compares relationship densities in given clusters to densities between clusters. The measure of the quality of those groupings is called modularity.

Modularity algorithms optimize communities locally and then globally, using multiple iterations to test different groupings and increasing coarseness. This strategy identifies community hierarchies and provides a broad understanding of the overall structure. However, all modularity algorithms suffer from two drawbacks:

- They merge smaller communities into larger ones.
- A plateau can occur where several partition options are present with similar modularity, forming local maxima and preventing progress.

For more information, see the paper "The Performance of Modularity Maximization in Practical Contexts" (*https://arxiv.org/abs/0910.0165*), by B. H. Good, Y.-A. de Montjoye, and A. Clauset.

Initially the Louvain Modularity algorithm optimizes modularity locally on all nodes, which finds small communities; then each small community is grouped into a larger conglomerate node and the first step is repeated until we reach a global optimum.

Calculating Modularity

A simple calculation of modularity is based on the fraction of the relationships within the given groups minus the expected fraction if relationships were distributed at random between all nodes. The value is always between 1 and –1, with positive values indicating more relationship density than you'd expect by chance and negative values indicating less density. Figure 6-11 illustrates several different modularity scores based on node groupings.

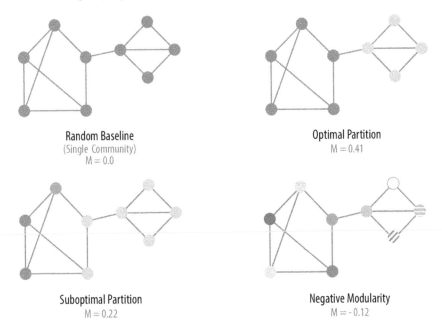

Figure 6-11. Four modularity scores based on different partitioning choices

The formula for the modularity of a group is:

$$M = \sum_{c=1}^{n_c} \left[\frac{L_c}{L} - \left(\frac{k_c}{2L} \right)^2 \right]$$

where:

- L is the number of relationships in the entire group.
- L_c is the number of relationships in a partition.
- k_c is the total degree of nodes in a partition.

The calculation for the optimal partition at the top of Figure 6-11 is as follows:

- The dark partition is $\left(\frac{7}{13} - \left(\frac{15}{2(13)}\right)^2\right) = 0.205$

- The light partition is $\left(\frac{5}{13} - \left(\frac{11}{2(13)}\right)^2\right) = 0.206$

- These are added together for $M = 0.205 + 0.206 = 0.41$

The algorithm consists of repeated application of two steps, as illustrated in Figure 6-12.

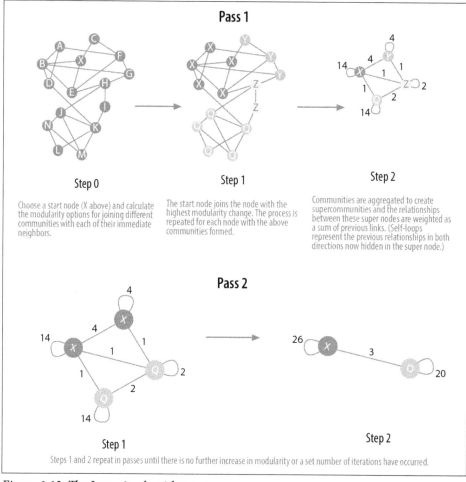

Figure 6-12. The Louvain algorithm process

The Louvain algorithm's steps include:

1. A "greedy" assignment of nodes to communities, favoring local optimizations of modularity.
2. The definition of a more coarse-grained network based on the communities found in the first step. This coarse-grained network will be used in the next iteration of the algorithm.

These two steps are repeated until no further modularity-increasing reassignments of communities are possible.

Part of the first optimization step is evaluating the modularity of a group. Louvain uses the following formula to accomplish this:

$$Q = \frac{1}{2m} \sum_{u,v} \left[A_{uv} - \frac{k_u k_v}{2m} \right] \delta(c_u, c_v)$$

where:

- u and v are nodes.
- m is the total relationship weight across the entire graph ($2m$ is a common normalization value in modularity formulas).
- $A_{uv} - \frac{k_u k_v}{2m}$ is the strength of the relationship between u and v compared to what we would expect with a random assignment (tends toward averages) of those nodes in the network.
  - A_{uv} is the weight of the relationship between u and v.
  - k_u is the sum of relationship weights for u.
  - k_v is the sum of relationship weights for v.
- $\delta(c_u, c_v)$ is equal to 1 if u and v are assigned to the same community, and 0 if they are not.

Another part of that first step evaluates the change in modularity if a node is moved to another group. Louvain uses a more complicated variation of this formula and then determines the best group assignment.

When Should I Use Louvain?

Use Louvain Modularity to find communities in vast networks. This algorithm applies a heuristic, as opposed to exact, modularity, which is computationally expensive. Louvain can therefore be used on large graphs where standard modularity algorithms may struggle.

Louvain is also very helpful for evaluating the structure of complex networks, in particular uncovering many levels of hierarchies–such as what you might find in a criminal organization. The algorithm can provide results where you can zoom in on different levels of granularity and find subcommunities within subcommunities within subcommunities.

Example use cases include:

- Detecting cyberattacks. The Louvain algorithm was used in a 2016 study by S. V. Shanbhaq (*https://bit.ly/2FAxalS*) of fast community detection in large-scale cybernetworks for cybersecurity applications. Once these communities have been detected they can be used to detect cyberattacks.

- Extracting topics from online social platforms, like Twitter and YouTube, based on the co-occurence of terms in documents as part of the topic modeling process. This approach is described in a paper by G. S. Kido, R. A. Igawa, and S. Barbon Jr., "Topic Modeling Based on Louvain Method in Online Social Networks" (*http://bit.ly/2UbCCUl*).

- Finding hierarchical community structures within the brain's functional network, as described in "Hierarchical Modularity in Human Brain Functional Networks" (*https://bit.ly/2HFHXxu*) by D. Meunier et al.

 Modularity optimization algorithms, including Louvain, suffer from two issues. First, the algorithms can overlook small communities within large networks. You can overcome this problem by reviewing the intermediate consolidation steps. Second, in large graphs with overlapping communities, modularity optimizers may not correctly determine the global maxima. In the latter case, we recommend using any modularity algorithm as a guide for gross estimation but not complete accuracy.

Louvain with Neo4j

Let's see the Louvain algorithm in action. Louvain takes in a config map with the following keys:

nodeProjection
> Enables the mapping of specific kinds of nodes into the in-memory graph. We can declare one or more node labels.

relationshipProjection
> Enables the mapping of relationship types into the in-memory graph. We can declare one or more relationship types along with direction and properties.

`includeIntermediateCommunities`

Indicates whether to write intermediate communities. If set to false, only the final community is persisted.

We can execute the following query to run the algorithm over our graph:

```
CALL gds.louvain.stream({
    nodeProjection: "Library",
    relationshipProjection: "DEPENDS_ON",
    includeIntermediateCommunities: true
})
YIELD nodeId, communityId, intermediateCommunityIds
RETURN gds.util.asNode(nodeId).id AS libraries,
       communityId, intermediateCommunityIds;
```

These are the results:

libraries	communityId	intermediateCommunityIds
"ipykernel"	0	[0, 0]
"jpy-client"	0	[2, 0]
"jpy-core"	0	[2, 0]
"six"	5	[3, 5]
"pandas"	5	[7, 5]
"numpy"	5	[5, 5]
"python-dateutil"	5	[3, 5]
"pytz"	5	[7, 5]
"pyspark"	11	[11, 11]
"matplotlib"	5	[3, 5]
"spacy"	5	[5, 5]
"py4j"	11	[11, 11]
"jupyter"	0	[2, 0]
"jpy-console"	0	[2, 0]
"nbconvert"	0	[2, 0]

The `intermediateCommunityIds` column describes the community that nodes fall into at two levels. The last value in the array is the final community and the other one is an intermediate community.

The numbers assigned to the intermediate and final communities are simply labels with no measurable meaning. Treat these as labels that indicate which community nodes belong to such as "belongs to a community labeled 0", "a community labeled 4", and so forth.

For example, matplotlib has a result of [3,5]. This means that matplotlib's final community is labeled 5 and its intermediate community is labeled 3.

The communityId column returns the final community for each node.

We can also store the final and intermediate communities using the write version of the algorithm and then query the results afterwards.

The following query will run the Louvain algorithm and store the list of communities in the communities property on each node:

```
CALL gds.louvain.write({
    nodeProjection: "Library",
    relationshipProjection: "DEPENDS_ON",
    includeIntermediateCommunities: true,
    writeProperty: "communities"
});
```

And the following query will run the Louvain algorithm and store the final community in the finalCommunity property on each node:

```
CALL gds.louvain.write({
    nodeProjection: "Library",
    relationshipProjection: "DEPENDS_ON",
    includeIntermediateCommunities: false,
    writeProperty: "finalCommunity"
});
```

Once we've run these two procedures, we can write the following query to explore the final clusters:

```
MATCH (l:Library)
RETURN l.finalCommunity AS community, collect(l.id) AS libraries
ORDER BY size(libraries) DESC;
```

Running the query yields this output:

community	libraries
5	["six", "pandas", "numpy", "python-dateutil", "pytz", "matplotlib", "spacy"]
0	["ipykernel", "jpy-client", "jpy-core", "jupyter", "jpy-console", "nbconvert"]
11	["pyspark", "py4j"]

This clustering is the same as we saw with the connected components algorithm.

matplotlib is in a community with pytz, spacy, six, pandas, numpy, and python-dateutil. We can see this more clearly in Figure 6-13.

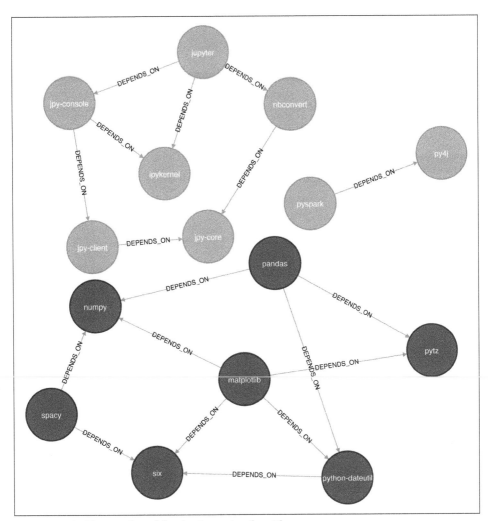

Figure 6-13. Clusters found by the Louvain algorithm

An additional feature of the Louvain algorithm is that we can see the intermediate clustering as well. The following query will show us finer-grained clusters than the final layer did:

```
MATCH (l:Library)
RETURN l.communities[0] AS community, collect(l.id) AS libraries
ORDER BY size(libraries) DESC;
```

Running that query gives this output:

community	libraries
2	["jpy-client", "jpy-core", "jupyter", "jpy-console", "nbconvert"]
3	["six", "python-dateutil", "matplotlib"]
7	["pandas", "pytz"]
5	["numpy", "spacy"]
11	["pyspark", "py4j"]
0	["ipykernel"]

The libraries in the matplotlib community have now broken down into three smaller communities:

- matplotlib, python-dateutil, and six
- pandas and pytz
- numpy and spacy

We can see this breakdown visually in Figure 6-14.

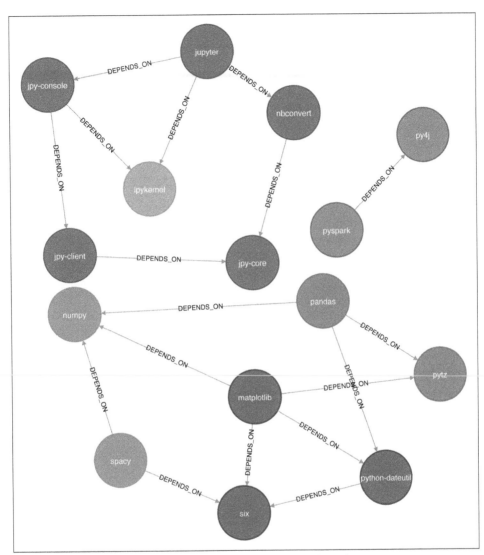

Figure 6-14. Intermediate clusters found by the Louvain algorithm

Although this graph only showed two layers of hierarchy, if we ran this algorithm on a larger graph we would see a more complex hierarchy. The intermediate clusters that Louvain reveals can be very useful for detecting fine-grained communities that may not be detected by other community detection algorithms.

Validating Communities

Community detection algorithms generally have the same goal: to identify groups. However, because different algorithms begin with different assumptions, they may uncover different communities. This makes choosing the right algorithm for a particular problem more challenging and a bit of an exploration.

Most community detection algorithms do reasonably well when relationship density is high within groups compared to their surroundings, but real-world networks are often less distinct. We can validate the accuracy of the communities found by comparing our results to a benchmark based on data with known communities.

Two of the best-known benchmarks are the Girvan-Newman (GN) and Lancichinetti–Fortunato–Radicchi (LFR) algorithms. The reference networks that these algorithms generate are quite different: GN generates a random network which is more homogeneous, whereas LFR creates a more heterogeneous graph where node degrees and community size are distributed according to a power law.

Since the accuracy of our testing depends on the benchmark used, it's important to match our benchmark to our dataset. As much as possible, look for similar densities, relationship distributions, community definitions, and related domains.

Summary

Community detection algorithms are useful for understanding the way that nodes are grouped together in a graph.

In this chapter, we started by learning about the Triangle Count and Clustering Coefficient algorithms. We then moved on to two deterministic community detection algorithms: Strongly Connected Components and Connected Components. These algorithms have strict definitions of what constitutes a community and are very useful for getting a feel for the graph structure early in the graph analytics pipeline.

We finished with Label Propagation and Louvain, two nondeterministic algorithms which are better able to detect finer-grained communities. Louvain also showed us a hierarchy of communities at different scales.

In the next chapter, we'll take a much larger dataset and learn how to combine the algorithms together to gain even more insight into our connected data.

CHAPTER 7

Graph Algorithms in Practice

The approach we take to graph analysis evolves as we become more familiar with the behavior of different algorithms on specific datasets. In this chapter, we'll run through several examples to give you a better feeling for how to tackle large-scale graph data analysis using datasets from Yelp and the US Department of Transportation. We'll walk through Yelp data analysis in Neo4j that includes a general overview of the data, combining algorithms to make trip recommendations, and mining user and business data for consulting. In Spark, we'll look into US airline data to understand traffic patterns and delays as well as how airports are connected by different airlines.

Because pathfinding algorithms are straightforward, our examples will use these centrality and community detection algorithms:

- PageRank to find influential Yelp reviewers and then correlate their ratings for specific hotels

- Betweenness Centrality to uncover reviewers connected to multiple groups and then extract their preferences

- Label Propagation with a projection to create supercategories of similar Yelp businesses

- Degree Centrality to quickly identify airport hubs in the US transport dataset

- Strongly Connected Components to look at clusters of airport routes in the US

Analyzing Yelp Data with Neo4j

Yelp helps people find local businesses based on reviews, preferences, and recommendations. Over 180 million reviews had been written on the platform as of the end of 2018. Since 2013, Yelp has run the Yelp Dataset challenge (*https://bit.ly/3fCL6vG*), a competition that encourages people to explore and research Yelp's open dataset.

As of Round 12 (conducted in 2018) of the challenge, the open dataset contained:

- Over 7 million reviews plus tips
- Over 1.5 million users and 280,000 pictures
- Over 188,000 businesses with 1.4 million attributes
- 10 metropolitan areas

Since its launch, the dataset has become popular, with hundreds of academic papers (*https://bit.ly/2upiaRz*) written using this material. The Yelp dataset represents real data that is very well structured and highly interconnected. It's a great showcase for graph algorithms that you can also download and explore.

> Yelp makes a subset of their data available for personal, educational, and academic purposes. The Yelp dataset is periodically updated which may make it necessary for you to alter how you load updated data and will likely alter some algorithm results.

Yelp Social Network

As well as writing and reading reviews about businesses, users of Yelp form a social network. Users can send friend requests to other users they've come across while browsing Yelp.com, or they can connect their address books or Facebook graphs.

The Yelp dataset also includes a social network. Figure 7-1 is a screen capture of the Friends section of Mark's Yelp profile.

Apart from the fact that Mark needs a few more friends, we're ready to start. To illustrate how we might analyze Yelp data in Neo4j, we'll use a scenario where we work for a travel information business. We'll first explore the Yelp data, and then look at how to help people plan trips with our app. We will walk through finding good recommendations for places to stay and things to do in major cities like Las Vegas.

Another part of our business scenario will involve consulting to travel-destination businesses. In one example we'll help hotels identify influential visitors and then businesses that they should target for cross-promotion programs.

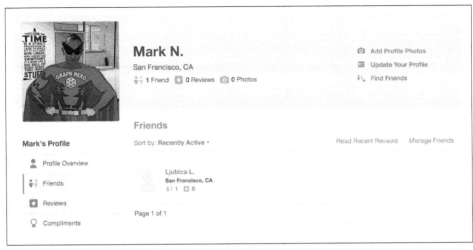

Figure 7-1. Mark's Yelp profile

Data Import

There are many different methods for importing data into Neo4j, including the Import tool (*https://bit.ly/2UTx26g*), the LOAD CSV command (*https://bit.ly/2CCfcgR*) that we've seen in earlier chapters, and Neo4j drivers (*https://bit.ly/2JDAr7U*).

For the Yelp dataset we need to do a one-off import of a large amount of data, so the Import tool is the best choice. See "Neo4j Bulk Data Import and Yelp" on page 237 for more details.

Graph Model

The Yelp data is represented in a graph model as shown in Figure 7-2.

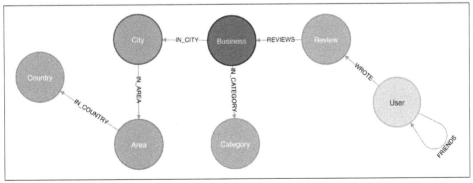

Figure 7-2. The Yelp graph model

Our graph contains User labeled nodes, which have FRIENDS relationships with other Users. Users also write Reviews and tips about Businesses. All of the metadata is stored as properties of nodes, except for business categories, which are represented by separate Category nodes. For location data we've extracted City, Area, and Country attributes into the subgraph. In other use cases it might make sense to extract other attributes to nodes such as dates, or collapse nodes to relationships such as reviews.

The Yelp dataset also includes user tips and photos, but we won't use those in our example.

A Quick Overview of the Yelp Data

Once we have the data loaded in Neo4j, we'll execute some exploratory queries. We'll ask how many nodes are in each category or what types of relations exist, to get a feel for the Yelp data. Previously we've shown Cypher queries for our Neo4j examples, but we might be executing these from another programming language. As Python is the go-to language for data scientists, we'll use Neo4j's Python driver in this section when we want to connect the results to other libraries from the Python ecosystem. If we just want to show the result of a query we'll use Cypher directly.

We'll also show how to combine Neo4j with the popular pandas library, which is effective for data wrangling outside of the database. We'll see how to use the tabulate library to prettify the results we get from pandas, and how to create visual representations of data using matplotlib.

We'll also be using Neo4j's APOC library of procedures to help us write even more powerful Cypher queries. There's more information about APOC in "APOC and Other Neo4j Tools" on page 239.

Let's first install the Python libraries:

```
pip install neo4j tabulate pandas matplotlib
```

Once we've done that we'll import those libraries:

```
from neo4j import GraphDatabase
import pandas as pd
from tabulate import tabulate
```

Importing matplotlib can be fiddly on macOS, but the following lines should do the trick:

```
import matplotlib
matplotlib.use('TkAgg')
import matplotlib.pyplot as plt
```

If you're running on another operating system, the middle line may not be required. Now let's create an instance of the Neo4j driver pointing at a local Neo4j database:

```
driver = GraphDatabase.driver("bolt://localhost", auth=("neo4j", "neo"))
```

You'll need to update the initialization of the driver to use your own host and credentials.

To get started, let's look at some general numbers for nodes and relationships. The following code calculates the cardinalities of node labels (i.e., counts the number of nodes for each label) in the database:

```
result = {"label": [], "count": []}
with driver.session() as session:
    labels = [row["label"] for row in session.run("CALL db.labels()")]
    for label in labels:
        query = f"MATCH (:`{label}`) RETURN count(*) as count"
        count = session.run(query).single()["count"]
        result["label"].append(label)
        result["count"].append(count)

df = pd.DataFrame(data=result)
print(tabulate(df.sort_values("count"), headers='keys',
                              tablefmt='psql', showindex=False))
```

If we run that code we'll see how many nodes we have for each label:

label	count
Country	17
Area	54
City	1093
Category	1293
Business	174567
User	1326101
Review	5261669

We could also create a visual representation of the cardinalities, with the following code:

```
plt.style.use('fivethirtyeight')

ax = df.plot(kind='bar', x='label', y='count', legend=None)

ax.xaxis.set_label_text("")
plt.yscale("log")
plt.xticks(rotation=45)
plt.tight_layout()
plt.show()
```

We can see the chart that gets generated by this code in Figure 7-3. Note that this chart is using log scale.

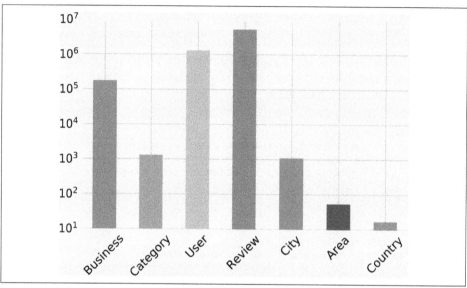

Figure 7-3. The number of nodes for each label category

Similarly, we can calculate the cardinalities of relationships:

```
result = {"relType": [], "count": []}
with driver.session() as session:
    rel_types = [row["relationshipType"] for row in session.run
                            ("CALL db.relationshipTypes()")]
    for rel_type in rel_types:
        query = f"MATCH ()-[:`{rel_type}`]->() RETURN count(*) as count"
        count = session.run(query).single()["count"]
        result["relType"].append(rel_type)
        result["count"].append(count)

df = pd.DataFrame(data=result)
print(tabulate(df.sort_values("count"), headers='keys',
                        tablefmt='psql', showindex=False))
```

If we run that code we'll see the number of each type of relationship:

relType	count
IN_COUNTRY	54
IN_AREA	1154
IN_CITY	174566
IN_CATEGORY	667527
WROTE	5261669
REVIEWS	5261669
FRIENDS	10645356

We can see a chart of the cardinalities in Figure 7-4. As with the node cardinalities chart, this chart is using log scale.

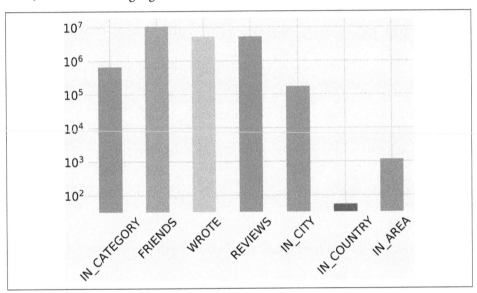

Figure 7-4. The number of relationships by relationship type

These queries shouldn't reveal anything surprising, but they're useful to get a feel for what's in the data. This also serves as a quick check that the data imported correctly.

We assume Yelp has many hotel reviews, but it makes sense to check before we focus on that sector. We can find out how many hotel businesses are in that data and how many reviews they have by running the following query:

```
MATCH (category:Category {name: "Hotels"})
RETURN size((category)<-[:IN_CATEGORY]-()) AS businesses,
       size((:Review)-[:REVIEWS]->(:Business)-[:IN_CATEGORY]->
                            (category)) AS reviews
```

Here's the result:

businesses	reviews
2683	183759

We have many businesses to work with, and a lot of reviews! In the next section we'll explore the data further with our business scenario.

Trip Planning App

To add well-liked recommendations to our app, we start by finding the most-rated hotels as a heuristic for popular choices for reservations. We can add how well they've been rated to understand the actual experience. To see the 10 most-reviewed hotels and plot their rating distributions, we use the following code:

```
query = """ ❶
MATCH (review:Review)-[:REVIEWS]->(business:Business),
      (business)-[:IN_CATEGORY]->(category:Category {name: $category}),
      (business)-[:IN_CITY]->(:City {name: $city})
RETURN business.name AS business, collect(review.stars) AS allReviews
ORDER BY size(allReviews) DESC
LIMIT 10
"""

fig = plt.figure()
fig.set_size_inches(10.5, 14.5)
fig.subplots_adjust(hspace=0.4, wspace=0.4)

with driver.session() as session:
    params = { "city": "Las Vegas", "category": "Hotels"}
    result = session.run(query, params)
    for index, row in enumerate(result):
        business = row["business"]
        stars = pd.Series(row["allReviews"])

        total = stars.count()
        average_stars = stars.mean().round(2)

        stars_histogram = stars.value_counts().sort_index() ❷
        stars_histogram /= float(stars_histogram.sum())
```

```
ax = fig.add_subplot(5, 2, index+1) ❸
stars_histogram.plot(kind="bar", legend=None, color="darkblue",
                     title=f"{business}\nAve:
                             {average_stars}, Total: {total}")

plt.tight_layout()
plt.show()
```

❶ Find the 10 hotels with the most reviews.

❷ Calculate the star distribution.

❸ Plot a bar chart showing the distribution of star ratings.

We've constrained by city and category to focus on Las Vegas hotels. We run that code we get the chart in Figure 7-5. Note that the x-axis represents the hotel's star rating and the y-axis represents the overall percentage of each rating.

These hotels have more reviews than anyone would likely read. It would be better to show users relevant reviews and make them more prominent on our app. For this analysis, we'll move from basic graph exploration to using graph algorithms.

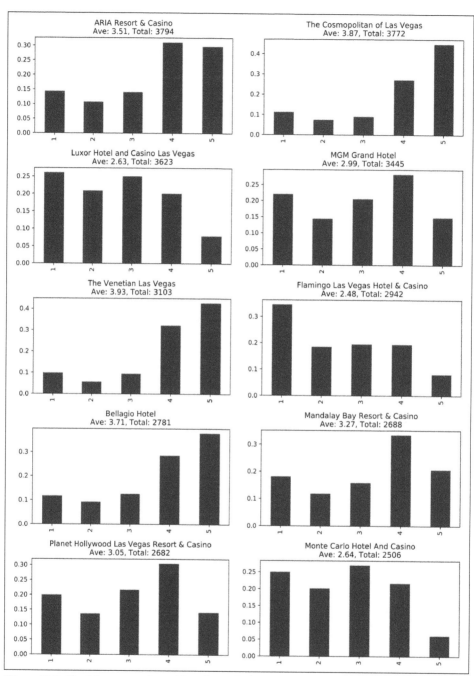

Figure 7-5. The 10 most-reviewed hotels, with the number of stars on the x-axis and their overall rating percentage on the y-axis

Finding influential hotel reviewers

One way we can decide which reviews to feature is by ordering reviews based on the *influence of the reviewer* on Yelp. We'll run the PageRank algorithm over the projected graph of all users that have reviewed at least three hotels. Remember from earlier chapters that a projection can help filter out inessential information as well as add relationship data (sometimes inferred). We'll use Yelp's friend graph (introduced in "Yelp Social Network" on page 154) as the relationships between users. The PageRank algorithm will uncover those reviewers with more sway over more users, even if they are not direct friends.

> If two people are Yelp friends there are two FRIENDS relationships between them. For example, if A and B are friends there will be a FRIENDS relationship from A to B and another from B to A.

We need to write a query that projects a subgraph of users with more than three reviews and then executes the PageRank algorithm over that projected subgraph.

It's easier to understand how the subgraph projection works with a small example. Figure 7-6 shows a graph of three mutual friends—Mark, Arya, and Praveena. Mark and Praveena have both reviewed three hotels and will be part of the projected graph. Arya, on the other hand, has only reviewed one hotel and will therefore be excluded from the projection.

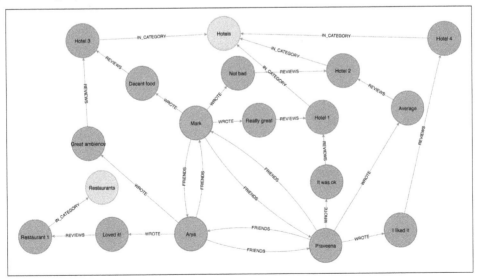

Figure 7-6. A sample Yelp graph

Our projected graph will only include Mark and Praveena, as shown in Figure 7-7.

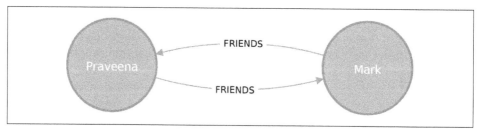

Figure 7-7. Our sample projected graph

Now that we've seen how graph projections work, let's move forward. The following query executes the PageRank algorithm over our projected graph and stores the result in the `hotelPageRank` property on each node:

```
CALL gds.pageRank.write({
    nodeQuery: 'MATCH (u:User)-[:WROTE]->()-[:REVIEWS]->(b),
                    (b)-[:IN_CATEGORY]->(:Category {name: $category})
                WITH u, count(*) AS reviews
                WHERE reviews >= $cutOff
                RETURN id(u) AS id',
    relationshipQuery: 'MATCH (u1:User)-[:WROTE]->()-[:REVIEWS]->(b),
                    (b)-[:IN_CATEGORY]->(:Category {name: $category})
                MATCH (u1)-[:FRIENDS]->(u2)
                RETURN id(u1) AS source, id(u2) AS target',
    writeProperty: "hotelPageRank",
    validateRelationships: false,
    parameters: {category: "Hotels", cutOff: 3}
})
YIELD nodePropertiesWritten, createMillis, computeMillis,
    writeMillis, ranIterations
RETURN nodePropertiesWritten, createMillis, computeMillis,
    writeMillis, ranIterations;
```

You might have noticed that we didn't set the damping factor or maximum iteration limit discussed in Chapter 5. If not explicitly set, Neo4j defaults to a `0.85` damping factor with `maxIterations` set to 20`.

If we run this query, we'll see the following output:

nodePropertiesWritten	createMillis	computeMillis	writeMillis	ranIterations
10195	7507	610	168	20

The `hotelPageRank` has been added to just over 10,000 users. Now let's look at the distribution of the PageRank values so we'll know how to filter our data:

```
MATCH (u:User)
WHERE exists(u.hotelPageRank)
RETURN count(u.hotelPageRank) AS count,
    avg(u.hotelPageRank) AS ave,
```

```
apoc.math.round(percentileDisc(u.hotelPageRank, 0.5), 2) AS `50%`,
apoc.math.round(percentileDisc(u.hotelPageRank, 0.75), 2) AS `75%`,
apoc.math.round(percentileDisc(u.hotelPageRank, 0.90), 2) AS `90%`,
apoc.math.round(percentileDisc(u.hotelPageRank, 0.95), 2) AS `95%`,
apoc.math.round(percentileDisc(u.hotelPageRank, 0.99), 2) AS `99%`,
apoc.math.round(percentileDisc(u.hotelPageRank, 0.999), 2) AS `99.9%`,
apoc.math.round(percentileDisc(u.hotelPageRank, 0.9999), 2) AS `99.99%`,
apoc.math.round(percentileDisc(u.hotelPageRank, 0.99999), 2) AS `99.999%`,
apoc.math.round(percentileDisc(u.hotelPageRank, 1), 2) AS `100%`;
```

If we run that query we'll get this output:

count	ave	50%	75%	90%	95%	99%	99.9%	99.99%	99.999%	100%
10195	0.7513906938612781	0.27	0.65	1.67	2.8	7.71	17.48	25.1	26.26	26.26

To interpret this percentile table, the 90% value of 1.67 means that 90% of users had a lower PageRank score. 99.99% reflects the influence rank for the top 0.01% of reviewers and 100% is simply the highest PageRank score.

It's interesting that 50% of the users have a score of under 0.27, which is slightly lower than the overall average—and only marginally more than the 0.15 that they are initialized with by the PageRank algorithm. It seems like this data reflects a power-law distribution with a few very influential reviewers.

Because we want to find only the most influential users, we'll write a query based on the PageRank score of the top 1,000 users. The following query finds reviewers with a PageRank score *higher* than 1.67 (notice that's the 99% group):

```
MATCH (u:User)
WHERE u.hotelPageRank > 1.67 ❶

WITH u ORDER BY u.hotelPageRank DESC ❷
LIMIT 10

RETURN u.name AS name,
       u.hotelPageRank AS pageRank,
       size((u)-[:WROTE]->()-[:REVIEWS]->()-[:IN_CATEGORY]->
           (:Category {name: "Hotels"})) AS hotelReviews,
       size((u)-[:WROTE]->()) AS totalReviews,
       size((u)-[:FRIENDS]-()) AS friends;
```

❶   Only find users that have a `hotelPageRank` score in the top 1,000 users.

❷   Find the top 10 of those users.

If we run that query we'll get the results seen here:

name	pageRank	hotelReviews	totalReviews	friends
Phil	26.258416866464536	15	134	8154
Philip	25.095720199332572	21	620	9634
Cassandra	19.118613678961992	3	23	7148
Carol	18.455617765896026	6	119	6218
Abby	18.014389160706198	9	82	7922
Joseph	18.008178111561577	5	32	6596
Erica	17.693355155712922	6	15	7071
J	17.663379741786045	103	1322	6498
Misti	17.638867230992766	19	730	6230
Randy	17.519849820295345	21	125	7846

These results show us that Phil is the most credible reviewer, although he hasn't reviewed many hotels. He's likely connected to some very influential people, but if we wanted a stream of new reviews, his profile wouldn't be the best selection. Philip has a slightly lower score, but has the most friends and has written five times more reviews than Phil. While J has written the most reviews of all and has a reasonable number of friends, J's PageRank score isn't the highest—but it's still in the top 10. For our app we choose to highlight hotel reviews from Phil, Philip, and J to give us the right mix of influencers and number of reviews.

Now that we've improved our in-app recommendations with relevant reviews, let's turn to the other side of our business: consulting.

Travel Business Consulting

As part of our consulting service, hotels subscribe to be alerted when an influential visitor writes about their stay so they can take any necessary action. First, we'll look at ratings of the Bellagio, sorted by the most influential reviewers:

```
query = """\
MATCH (b:Business {name: $hotel})
MATCH (b)<-[:REVIEWS]-(review)<-[:WROTE]-(user)
WHERE exists(user.hotelPageRank)
RETURN user.name AS name,
       user.hotelPageRank AS pageRank,
       review.stars AS stars
"""

with driver.session() as session:
    params = { "hotel": "Bellagio Hotel" }
    df = pd.DataFrame([dict(record) for record in session.run(query, params)])
    df = df.round(2)
    df = df[["name", "pageRank", "stars"]]
```

```
top_reviews = df.sort_values(by=["pageRank"], ascending=False).head(10)
print(tabulate(top_reviews, headers='keys', tablefmt='psql', showindex=False))
```

If we run that code we'll get this output:

name	pageRank	stars
Erica	17.69	4
J	17.66	5
Misti	17.64	5
Michael	17.37	4
Christine	15.84	4
Jeremy	14.94	5
Connie	14.6	5
Joyce	11.76	4
Henry	10.89	5
Flora	10.34	4

Note that these results are different from our previous table of the best hotel reviewers. That's because here we are only looking at reviewers that have rated the Bellagio.

Things are looking good for the hotel customer service team at the Bellagio—the top 10 influencers all give their hotel good rankings. They may want to encourage these people to visit again and share their experiences.

Are there any influential guests who haven't had such a good experience? We can run the following code to find the guests with the highest PageRank that rated their experience with fewer than four stars:

```
query = """\
MATCH (b:Business {name: $hotel})
MATCH (b)<-[:REVIEWS]-(review)<-[:WROTE]-(user)
WHERE exists(user.hotelPageRank) AND review.stars < $goodRating
RETURN user.name AS name,
       user.hotelPageRank AS pageRank,
       review.stars AS stars
"""

with driver.session() as session:
    params = { "hotel": "Bellagio Hotel", "goodRating": 4 }
    df = pd.DataFrame([dict(record) for record in session.run(query, params)])
    df = df.round(2)
    df = df[["name", "pageRank", "stars"]]

top_reviews = df.sort_values(by=["pageRank"], ascending=False).head(10)
print(tabulate(top_reviews, headers='keys', tablefmt='psql', showindex=False))
```

If we run that code we'll get the following results:

name	pageRank	stars
Chris	8.89	3
Lorrie	7.46	2
Victor	5.66	3
Dani	5.54	1
Francine	4.3	3
Rex	3.96	2
Jon	3.77	3
Stephen	3.75	2
Pasquale	3.6	2
Rachel	3.58	3

Our highest-ranked users giving the Bellagio lower ratings, Chris and Lorrie, are amongst the top 1,000 most influential users (as per the results of our earlier query), so perhaps a personal outreach is warranted. Also, because many reviewers write during their stay, real-time alerts about influencers may facilitate even more positive interactions.

Bellagio cross-promotion

After we helped them find influential reviewers, the Bellagio has now asked us to help identify other businesses for cross-promotion with help from well-connected customers. In our scenario, we recommend that they increase their customer base by attracting new guests from different types of communities as a greenfield opportunity. We can use the Betweenness Centrality algorithm that we discussed earlier to work out which Bellagio reviewers are not only well connected across the whole Yelp network, but might also act as a bridge between different groups.

We're only interested in finding influencers in Las Vegas, so we'll first tag those users:

```
MATCH (u:User)
WHERE exists((u)-[:WROTE]->()-[:REVIEWS]->()-[:IN_CITY]->
                              (:City {name: "Las Vegas"}))
SET u:LasVegas;
```

It would take a long time to run the Betweenness Centrality algorithm over our Las Vegas users, so instead we'll use the the RA-Brandes variant. This algorithm calculates a betweenness score by sampling nodes and only exploring shortest paths to a certain depth.

After some experimentation, we improved results with a few parameters set differently than the default values. We'll use shortest paths of up to 4 hops (maxDepth of 4) and sample 20% of the nodes (probability of 0.2). Note that increasing the number of hops and nodes will generally increase the accuracy, but at the cost of more time to compute the results. For any particular problem, the optimal parameters typically require testing to identify the point of diminishing returns.

The following query will execute the algorithm and store the result in the `between` property:

```
CALL gds.alpha.betweenness.sampled.write({
  nodeProjection: "LasVegas",
  relationshipProjection: "FRIENDS",
  maxDepth: 4,
  probability: 0.2,
  writeProperty: "between"
})
YIELD nodes, minCentrality, maxCentrality, sumCentrality
RETURN nodes, minCentrality, maxCentrality, sumCentrality;
```

This query will take a while to run, but once it's finished we'll see the following output:

nodes	minCentrality	maxCentrality	sumCentrality
506028	0.0	1980034772.4726186	162296447142.80072

Before we use these scores in our queries, let's write a quick exploratory query to see how the scores are distributed:

```
MATCH (u:User)
WHERE exists(u.between)
RETURN count(u.between) AS count,
       avg(u.between) AS ave,
       toInteger(percentileDisc(u.between, 0.5)) AS `50%`,
       toInteger(percentileDisc(u.between, 0.75)) AS `75%`,
       toInteger(percentileDisc(u.between, 0.90)) AS `90%`,
       toInteger(percentileDisc(u.between, 0.95)) AS `95%`,
       toInteger(percentileDisc(u.between, 0.99)) AS `99%`,
       toInteger(percentileDisc(u.between, 0.999)) AS `99.9%`,
       toInteger(percentileDisc(u.between, 0.9999)) AS `99.99%`,
       toInteger(percentileDisc(u.between, 1)) AS p100;
```

If we run that query we'll see the following output:

count	ave	50%	75%	90%	95%	99%	99.9%	99.99%	100%
506028	320726.21898944734	0	9960	316822	1010165	4418317	35294323	214059045	1980034772

Half of our users have a score of 0, meaning they are not well connected at all. The top 1 percentile (99% column) are on at least 4 million shortest paths between our set of 500,000 users. Considered together, we know that most of our users are poorly connected, but a few exert a lot of control over information; this is a classic behavior of small-world networks.

We can find out who our superconnectors are by running the following query:

```
MATCH (u:User)-[:WROTE]->()-[:REVIEWS]->(:Business {name:"Bellagio Hotel"})
WHERE exists(u.between)
RETURN u.name AS user,
       toInteger(u.between) AS betweenness,
       u.hotelPageRank AS pageRank,
       size((u)-[:WROTE]->()-[:REVIEWS]->()-[:IN_CATEGORY]->
                         (:Category {name: "Hotels"}))
           AS hotelReviews
ORDER BY u.between DESC
LIMIT 10;
```

The output is as follows:

user	betweenness	pageRank	hotelReviews
Misti	840712928	17.638867230992766	19
Christine	241898939	15.836424705456013	16
Erica	235666410	17.693355155712922	6
Mike	215417753	NULL	2
J	189802203	17.663379741786045	103
Michael	162077742	7.693667471793014	31
Jeremy	159790780	14.941242662875448	6
Michael	143599018	17.371501608518887	13
Chris	138279165	8.889553494018038	5
Connie	132470615	14.597890720120633	7

We see some of the same people here that we saw earlier in our PageRank query, with Mike being an interesting exception. He was excluded from that calculation because he hasn't reviewed enough hotels (three was the cutoff), but it seems like he's quite well connected in the world of Las Vegas Yelp users.

In an effort to reach a wider variety of customers, we'll look at other preferences these "connectors" display to see what we should promote. Many of these users have also reviewed restaurants, so we write the following query to find out which ones they like best:

```
MATCH (u:User)-[:WROTE]->()-[:REVIEWS]->(:Business {name:"Bellagio Hotel"})  ❶
WHERE u.between > 4418317
WITH u ORDER BY u.between DESC LIMIT 50

MATCH (u)-[:WROTE]->(review)-[:REVIEWS]-(business)  ❷
WHERE (business)-[:IN_CATEGORY]->(:Category {name: "Restaurants"})
AND   (business)-[:IN_CITY]->(:City {name: "Las Vegas"})

WITH business, avg(review.stars) AS averageReview, count(*) AS numberOfReviews  ❸
WHERE numberOfReviews >= 3
```

```
RETURN business.name AS business, averageReview, numberOfReviews
ORDER BY averageReview DESC, numberOfReviews DESC
LIMIT 10;
```

❶   Find the top 50 users who have reviewed the Bellagio

❷   Find the restaurants those users have reviewed in Las Vegas

❸   Only include restaurants that have more than 3 reviews by these users

This query finds our top 50 influential connectors, and finds the top 10 Las Vegas restaurants where at least 3 of them have rated the restaurant. If we run it, we'll see the
output shown here:

business	averageReview	numberOfReviews
Jean Georges Steakhouse	5.0	6
Sushi House Goyemon	5.0	6
Parma By Chef Marc	5.0	4
Kabuto	5.0	4
é by José Andrés	5.0	4
Yonaka Modern Japanese	5.0	4
Art of Flavors	5.0	4
Buffet of Buffets	5.0	3
Rose. Rabbit. Lie	5.0	3
Harvest by Roy Ellamar	5.0	3

We can now recommend that the Bellagio run a joint promotion with these restaurants to attract new guests from groups they might not typically reach. Superconnectors who rate the Bellagio well become our proxy for estimating which restaurants
might catch the eye of new types of target visitors.

Now that we have helped the Bellagio reach new groups, we're going to see how we
can use community detection to further improve our app.

Finding Similar Categories

While our end users are using the app to find hotels, we want to showcase other businesses they might be interested in. The Yelp dataset contains more than 1,000 categories, and it seems likely that some of those categories are similar to each other. We'll
use that similarity to make in-app recommendations for new businesses that our
users will likely find interesting.

Our graph model doesn't have any relationships between categories, but we can use the ideas described in "Monopartite, Bipartite, and k-Partite Graphs" on page 25 to build a category similarity graph based on how businesses categorize themselves.

For example, imagine that only one business categorizes itself under both Hotels and Historical Tours, as seen in Figure 7-8.

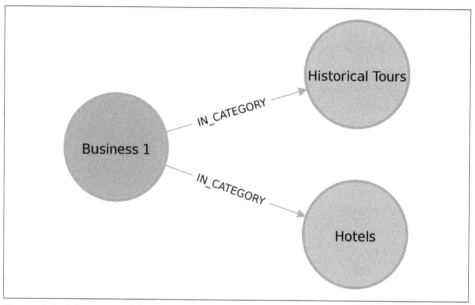

Figure 7-8. A business with two categories

This would result in a projected graph that has a link between Hotels and Historical Tours with a weight of 1, as seen in Figure 7-9.

Figure 7-9. A projected categories graph

In this case, we don't actually have to create the similarity graph—instead, we can run a community detection algorithm such as Label Propagation over a projected similarity graph. Using Label Propagation will effectively cluster businesses around the supercategory with which they have most in common:

```
CALL gds.labelPropagation.stream({
  nodeQuery: 'MATCH (c:Category) RETURN id(c) AS id',
  relationshipQuery: 'MATCH (c1:Category)<-[:IN_CATEGORY]-(b),
                            (c2:Category)<-[:IN_CATEGORY]-(b)
                      WHERE id(c1) < id(c2)
                      RETURN id(c1) AS source,
                             id(c2) AS target,
                             count(*) AS weight'
})
YIELD nodeId, communityId
WITH gds.util.asNode(nodeId) AS node, communityId
MERGE (sc:SuperCategory {name: "SuperCategory-" + communityId})
MERGE (node)-[:IN_SUPER_CATEGORY]->(sc);
```

Let's give those supercategories a friendlier name—the name of their largest category works well here:

```
MATCH (sc:SuperCategory)<-[:IN_SUPER_CATEGORY]-(category)
WITH sc, category, size((category)<-[:IN_CATEGORY]-()) as size
ORDER BY size DESC
WITH sc, collect(category.name)[0] as biggestCategory
SET sc.friendlyName = "SuperCat " + biggestCategory;
```

We can see a sample of categories and supercategories in Figure 7-10.

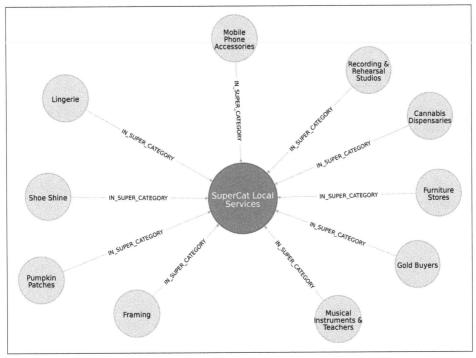

Figure 7-10. Categories and supercategories

The following query finds the most prevalent similar categories to Hotels in Las Vegas:

```
MATCH (hotels:Category {name: "Hotels"}),
      (lasVegas:City {name: "Las Vegas"}),
      (hotels)-[:IN_SUPER_CATEGORY]->(superCat),
      (superCat)<-[:IN_SUPER_CATEGORY]-(otherCategory)
RETURN otherCategory.name AS otherCategory,
       size((otherCategory)<-[:IN_CATEGORY]-(:Business)-
                           [:IN_CITY]->(lasVegas)) AS businesses
ORDER BY businesses DESC
LIMIT 10;
```

If we run that query we'll see the following output:

otherCategory	businesses
Tours	189
Car Rental	160
Limos	84
Resorts	73
Airport Shuttles	52
Taxis	35
Vacation Rentals	29
Airports	25
Airlines	23
Motorcycle Rental	19

Do these results seem odd? Obviously taxis and tours aren't hotels, but remember that this is based on self-reported categorizations. What the Label Propagation algorithm is really showing us in this similarity group are adjacent businesses and services.

Now let's find some businesses with an above-average rating in each of those categories:

```
MATCH (hotels:Category {name: "Hotels"}), ❶
      (hotels)-[:IN_SUPER_CATEGORY]->(superCat),
      (otherCategory)-[:IN_SUPER_CATEGORY]->(superCat),
      (otherCategory)<-[:IN_CATEGORY]-(business)
WHERE (business)-[:IN_CITY]->(:City {name: "Las Vegas"})

WITH otherCategory, count(*) AS count, ❷
     collect(business) AS businesses,
     percentileDisc(business.averageStars, 0.9) AS p90Stars
ORDER BY rand() DESC
LIMIT 10
```

```
WITH otherCategory,
     [b in businesses where b.averageStars >= p90Stars] AS businesses  ❸

WITH otherCategory,
     businesses[toInteger(rand() * size(businesses))] AS business  ❹

RETURN otherCategory.name AS otherCategory,
       business.name AS business,
       business.averageStars AS averageStars;
```

❶ Find businesses in Las Vegas that have the same SuperCategory as Hotels.

❷ Select 10 random categories and calculate the 90th percentile star rating.

❸ Select businesses from each of those categories that have an average rating higher than the 90th percentile using a pattern comprehension.

❹ Select one business per category.

In this query we use pattern comprehension (*https://bit.ly/2HRa1gr*) for the first time. Pattern comprehension is a syntax construction for creating a list based on pattern matching. It finds a specified pattern using a MATCH clause with a WHERE clause for predicates and then yields a custom projection. This Cypher feature was added based on inspiration from GraphQL (*https://graphql.org*), a query language for APIs.

If we run that query we see the following result:

otherCategory	business	averageStars
Motorcycle Rental	Adrenaline Rush Slingshot Rentals	5.0
Snorkeling	Sin City Scuba	5.0
Guest Houses	Hotel Del Kacvinsky	5.0
Car Rental	The Lead Team	5.0
Food Tours	Taste BUZZ Food Tours	5.0
Airports	Signature Flight Support	5.0
Public Transportation	JetSuiteX	4.6875
Ski Resorts	Trikke Las Vegas	4.833333333333332
Town Car Service	MW Travel Vegas	4.866666666666665
Campgrounds	McWilliams Campground	3.875

We can then make real-time recommendations based on a user's immediate app behavior. For example, while users are looking at Las Vegas hotels, we can now highlight a variety of adjacent Las Vegas businesses with good ratings. We can generalize these approaches to any business category, such as restaurants or theaters, in any location.

Reader Exercises

- Can you plot how the reviews for a city's hotels vary over time?
- What about for a particular hotel or other business?
- Are there any trends (seasonal or otherwise) in popularity?
- Do the most influential reviewers connect (out-link) to only other influential reviewers?

Analyzing Airline Flight Data with Apache Spark

In this section, we'll use a different scenario to illustrate the analysis of US airport data with Spark. Imagine you're a data scientist with a considerable travel schedule and would like to dig into information about airline flights and delays. We'll first explore airport and flight information and then look deeper into delays at two specific airports. Community detection will be used to analyze routes and find the best use of our frequent flyer points.

The US Bureau of Transportation Statistics makes available a significant amount of transportation information (*https://bit.ly/2Fy0GHA*). For our analysis, we'll use their May 2018 air travel on-time performance data, which includes flights originating and ending in the United States in that month. To add more detail about airports, such as location information, we'll also load data from a separate source, OpenFlights (*https://bit.ly/2Frv8TO*).

Let's load the data in Spark. As was the case in previous sections, our data is in CSV files that are available on the book's Github repository (*https://bit.ly/2FPgGVV*).

```
nodes = spark.read.csv("data/airports.csv", header=False)

cleaned_nodes = (nodes.select("_c1", "_c3", "_c4", "_c6", "_c7")
                .filter("_c3 = 'United States'")
                .withColumnRenamed("_c1", "name")
                .withColumnRenamed("_c4", "id")
                .withColumnRenamed("_c6", "latitude")
                .withColumnRenamed("_c7", "longitude")
                .drop("_c3"))
cleaned_nodes = cleaned_nodes[cleaned_nodes["id"] != "\\N"]
```

```
relationships = spark.read.csv("data/188591317_T_ONTIME.csv", header=True)

cleaned_relationships = (relationships
                        .select("ORIGIN", "DEST", "FL_DATE", "DEP_DELAY",
                                "ARR_DELAY", "DISTANCE", "TAIL_NUM", "FL_NUM",
                                "CRS_DEP_TIME", "CRS_ARR_TIME",
                                "UNIQUE_CARRIER")
                        .withColumnRenamed("ORIGIN", "src")
                        .withColumnRenamed("DEST", "dst")
                        .withColumnRenamed("DEP_DELAY", "deptDelay")
                        .withColumnRenamed("ARR_DELAY", "arrDelay")
                        .withColumnRenamed("TAIL_NUM", "tailNumber")
                        .withColumnRenamed("FL_NUM", "flightNumber")
                        .withColumnRenamed("FL_DATE", "date")
                        .withColumnRenamed("CRS_DEP_TIME", "time")
                        .withColumnRenamed("CRS_ARR_TIME", "arrivalTime")
                        .withColumnRenamed("DISTANCE", "distance")
                        .withColumnRenamed("UNIQUE_CARRIER", "airline")
                        .withColumn("deptDelay",
                            F.col("deptDelay").cast(FloatType()))
                        .withColumn("arrDelay",
                            F.col("arrDelay").cast(FloatType()))
                        .withColumn("time", F.col("time").cast(IntegerType()))
                        .withColumn("arrivalTime",
                            F.col("arrivalTime").cast(IntegerType()))
                        )

g = GraphFrame(cleaned_nodes, cleaned_relationships)
```

We have to do some cleanup on the nodes because some airports don't have valid air‐
port codes. We'll give the columns more descriptive names and convert some items
into appropriate numeric types. We also need to make sure that we have columns
named id, dst, and src, as this is expected by Spark's GraphFrames library.

We'll also create a separate DataFrame that maps airline codes to airline names. We'll
use this later in this chapter:

```
airlines_reference = (spark.read.csv("data/airlines.csv")
        .select("_c1", "_c3")
        .withColumnRenamed("_c1", "name")
        .withColumnRenamed("_c3", "code"))

airlines_reference = airlines_reference[airlines_reference["code"] != "null"]
```

Exploratory Analysis

Let's start with some exploratory analysis to see what the data looks like.

First let's see how many airports we have:

```
g.vertices.count()
```

```
1435
```

And how many connections do we have between these airports?

```
g.edges.count()
```

```
616529
```

Popular Airports

Which airports have the most departing flights? We can work out the number of out-going flights from an airport using the Degree Centrality algorithm:

```
airports_degree = g.outDegrees.withColumnRenamed("id", "oId")

full_airports_degree = (airports_degree
                        .join(g.vertices, airports_degree.oId == g.vertices.id)
                        .sort("outDegree", ascending=False)
                        .select("id", "name", "outDegree"))

full_airports_degree.show(n=10, truncate=False)
```

If we run that code we'll see the following output:

id	name	outDegree
ATL	Hartsfield Jackson Atlanta International Airport	33837
ORD	Chicago O'Hare International Airport	28338
DFW	Dallas Fort Worth International Airport	23765
CLT	Charlotte Douglas International Airport	20251
DEN	Denver International Airport	19836
LAX	Los Angeles International Airport	19059
PHX	Phoenix Sky Harbor International Airport	15103
SFO	San Francisco International Airport	14934
LGA	La Guardia Airport	14709
IAH	George Bush Intercontinental Houston Airport	14407

Most large US cities show up on this list—Chicago, Atlanta, Los Angeles, and New York all have popular airports. We can also create a visual representation of the outgoing flights using the following code:

```
plt.style.use('fivethirtyeight')

ax = (full_airports_degree
      .toPandas()
      .head(10)
      .plot(kind='bar', x='id', y='outDegree', legend=None))

ax.xaxis.set_label_text("")
plt.xticks(rotation=45)
plt.tight_layout()
plt.show()
```

The resulting chart can be seen in Figure 7-11.

It's quite striking how suddenly the number of flights drops off. Denver International Airport (DEN), the fifth most popular airport, has just over half as many outgoing fights as Hartsfield Jackson Atlanta International Airport (ATL), in first place.

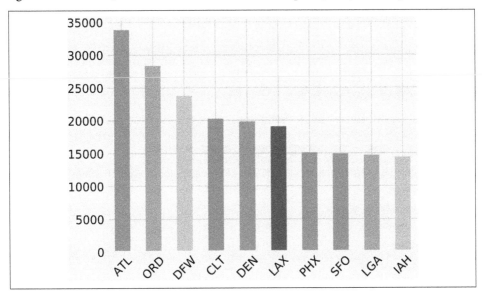

Figure 7-11. Outgoing flights by airport

Delays from ORD

In our scenario, we frequently travel between the west and east coasts and want to see delays through a midpoint hub like Chicago O'Hare International Airport (ORD). This dataset contains flight delay data, so we can dive right in.

The following code finds the average delay of flights departing from ORD grouped by the destination airport:

```
delayed_flights = (g.edges
                    .filter("src = 'ORD' and deptDelay > 0")
                    .groupBy("dst")
                    .agg(F.avg("deptDelay"), F.count("deptDelay"))
                    .withColumn("averageDelay",
                                F.round(F.col("avg(deptDelay)"), 2))
                    .withColumn("numberOfDelays",
                                F.col("count(deptDelay)")))

(delayed_flights
 .join(g.vertices, delayed_flights.dst == g.vertices.id)
 .sort(F.desc("averageDelay"))
 .select("dst", "name", "averageDelay", "numberOfDelays")
 .show(n=10, truncate=False))
```

Once we've calculated the average delay grouped by destination we join the resulting Spark DataFrame with a DataFrame containing all vertices, so that we can print the full name of the destination airport.

Running this code returns the 10 destinations with the worst delays:

dst	name	averageDelay	numberOfDelays
CKB	North Central West Virginia Airport	145.08	12
OGG	Kahului Airport	119.67	9
MQT	Sawyer International Airport	114.75	12
MOB	Mobile Regional Airport	102.2	10
TTN	Trenton Mercer Airport	101.18	17
AVL	Asheville Regional Airport	98.5	28
ISP	Long Island Mac Arthur Airport	94.08	13
ANC	Ted Stevens Anchorage International Airport	83.74	23
BTV	Burlington International Airport	83.2	25
CMX	Houghton County Memorial Airport	79.18	17

This is interesting, but one data point really stands out: 12 flights from ORD to CKB have been delayed by more than 2 hours on average! Let's find the flights between those airports and see what's going on:

```
from_expr = 'id = "ORD"'
to_expr = 'id = "CKB"'
ord_to_ckb = g.bfs(from_expr, to_expr)

ord_to_ckb = ord_to_ckb.select(
  F.col("e0.date"),
  F.col("e0.time"),
  F.col("e0.flightNumber"),
  F.col("e0.deptDelay"))
```

We can then plot the flights with the following code:

```
ax = (ord_to_ckb
      .sort("date")
      .toPandas()
      .plot(kind='bar', x='date', y='deptDelay', legend=None))

ax.xaxis.set_label_text("")
plt.tight_layout()
plt.show()
```

If we run that code we'll get the chart in Figure 7-12.

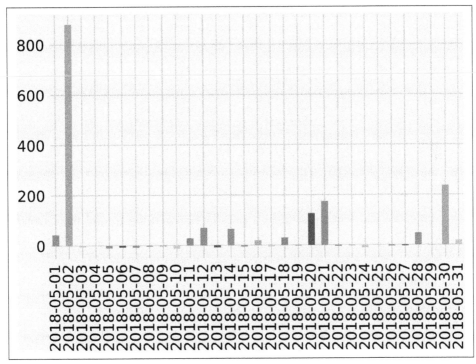

Figure 7-12. Flights from ORD to CKB

About half of the flights were delayed, but the delay of more than 14 hours on May 2, 2018, has massively skewed the average.

What if we want to find delays coming into and going out of a coastal airport? Those airports are often affected by adverse weather conditions, so we might be able to find some interesting delays.

Bad Day at SFO

Let's consider delays at an airport known for fog-related "low ceiling" issues: San Francisco International Airport (SFO). One method for analysis would be to look at *motifs*, which are recurrent subgraphs or patterns.

 The equivalent to motifs in Neo4j is graph patterns, which are found using the MATCH clause or with pattern expressions in Cypher.

GraphFrames lets us search for motifs (*http://bit.ly/2TZQ89B*), so we can use the structure of flights as part of a query. Let's use motifs to find the most-delayed flights going into and out of SFO on May 11, 2018. The following code will find these delays:

```
motifs = (g.find("(a)-[ab]->(b); (b)-[bc]->(c)")
          .filter("""(b.id = 'SFO') and
              (ab.date = '2018-05-11' and bc.date = '2018-05-11') and
              (ab.arrDelay > 30 or bc.deptDelay > 30) and
              (ab.flightNumber = bc.flightNumber) and
              (ab.airline = bc.airline) and
              (ab.time < bc.time)"""))
```

The motif (a)-[ab]->(b); (b)-[bc]->(c) finds flights coming into and out from the same airport. We then filter the resulting pattern to find flights with:

- A sequence where the first flight arrives at SFO and the second flight departs from SFO
- Delays of over 30 minutes when arriving at *or* departing from SFO
- The same flight number and airline

We can then take the result and select the columns we're interested in:

```
result = (motifs.withColumn("delta", motifs.bc.deptDelay - motifs.ab.arrDelay)
          .select("ab", "bc", "delta")
          .sort("delta", ascending=False))

result.select(
    F.col("ab.src").alias("a1"),
    F.col("ab.time").alias("a1DeptTime"),
```

```
        F.col("ab.arrDelay"),
        F.col("ab.dst").alias("a2"),
        F.col("bc.time").alias("a2DeptTime"),
        F.col("bc.deptDelay"),
        F.col("bc.dst").alias("a3"),
        F.col("ab.airline"),
        F.col("ab.flightNumber"),
        F.col("delta")
    ).show()
```

We're also calculating the *delta* between the arriving and departing flights to see which delays we can truly attribute to SFO.

If we execute this code we'll get the following result:

airline	flightNumber	a1	a1DeptTime	arrDelay	a2	a2DeptTime	deptDelay	a3	delta
WN	1454	PDX	1130	-18.0	SFO	1350	178.0	BUR	196.0
OO	5700	ACV	1755	-9.0	SFO	2235	64.0	RDM	73.0
UA	753	BWI	700	-3.0	SFO	1125	49.0	IAD	52.0
UA	1900	ATL	740	40.0	SFO	1110	77.0	SAN	37.0
WN	157	BUR	1405	25.0	SFO	1600	39.0	PDX	14.0
DL	745	DTW	835	34.0	SFO	1135	44.0	DTW	10.0
WN	1783	DEN	1830	25.0	SFO	2045	33.0	BUR	8.0
WN	5789	PDX	1855	119.0	SFO	2120	117.0	DEN	-2.0
WN	1585	BUR	2025	31.0	SFO	2230	11.0	PHX	-20.0

The worst offender, WN 1454, is shown in the top row; it arrived early but departed almost three hours late. We can also see that there are some negative values in the arrDelay column; this means that the flight into SFO was early.

Also notice that some flights, such as WN 5789 and WN 1585, made up time while on the ground in SFO, as shown with a negative delta.

Interconnected Airports by Airline

Now let's say we've traveled a lot, and those frequent flyer points we're determined to use to see as many destinations as efficiently as possible are soon to expire. If we start from a specific US airport, how many different airports can we visit and come back to the starting airport using the same airline?

Let's first identify all the airlines and work out how many flights there are on each of them:

```
airlines = (g.edges
  .groupBy("airline")
  .agg(F.count("airline").alias("flights"))
  .sort("flights", ascending=False))

full_name_airlines = (airlines_reference
                      .join(airlines, airlines.airline
                            == airlines_reference.code)
                      .select("code", "name", "flights"))
```

And now let's create a bar chart showing our airlines:

```
ax = (full_name_airlines.toPandas()
      .plot(kind='bar', x='name', y='flights', legend=None))

ax.xaxis.set_label_text("")
plt.tight_layout()
plt.show()
```

If we run that query we'll get the output in Figure 7-13.

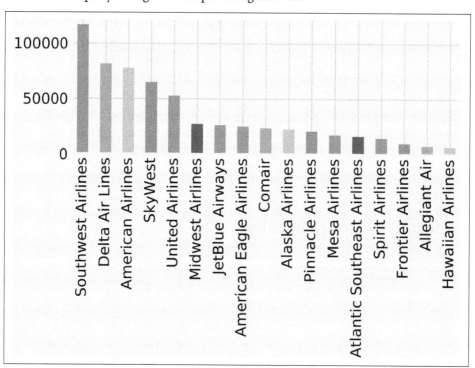

Figure 7-13. The number of flights by airline

Now let's write a function that uses the Strongly Connected Components algorithm to find airport groupings for each airline where all the airports have flights to and from all the other airports in that group:

```
def find_scc_components(g, airline):
    # Create a subgraph containing only flights on the provided airline
    airline_relationships = g.edges[g.edges.airline == airline]
    airline_graph = GraphFrame(g.vertices, airline_relationships)

    # Calculate the Strongly Connected Components
    scc = airline_graph.stronglyConnectedComponents(maxIter=10)

    # Find the size of the biggest component and return that
    return (scc
        .groupBy("component")
        .agg(F.count("id").alias("size"))
        .sort("size", ascending=False)
        .take(1)[0]["size"])
```

We can write the following code to create a DataFrame containing each airline and the number of airports of its largest strongly connected component:

```
# Calculate the largest strongly connected component for each airline
airline_scc = [(airline, find_scc_components(g, airline))
               for airline in airlines.toPandas()["airline"].tolist()]
airline_scc_df = spark.createDataFrame(airline_scc, ['id', 'sccCount'])

# Join the SCC DataFrame with the airlines DataFrame so that we can show
# the number of flights an airline has alongside the number of
# airports reachable in its biggest component
airline_reach = (airline_scc_df
    .join(full_name_airlines, full_name_airlines.code == airline_scc_df.id)
    .select("code", "name", "flights", "sccCount")
    .sort("sccCount", ascending=False))
```

And now let's create a bar chart showing our airlines:

```
ax = (airline_reach.toPandas()
      .plot(kind='bar', x='name', y='sccCount', legend=None))

ax.xaxis.set_label_text("")
plt.tight_layout()
plt.show()
```

If we run that query we'll get the output in Figure 7-14.

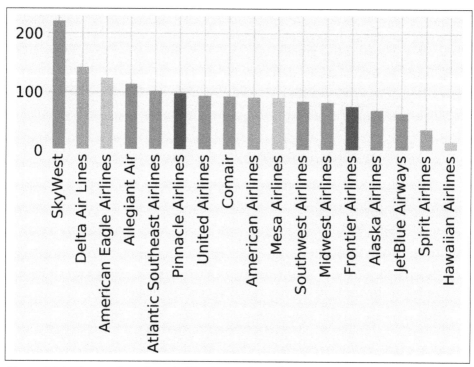

Figure 7-14. The number of reachable airports by airline

SkyWest has the largest community, with over 200 strongly connected airports. This might partially reflect its business model as an affiliate airline which operates aircraft used on flights for partner airlines. Southwest, on the other hand, has the highest number of flights but only connects around 80 airports.

Now let's say most of the frequent flyer points we have are with Delta Airlines (DL). Can we find airports that form communities within the network for that particular airline carrier?

```
airline_relationships = g.edges.filter("airline = 'DL'")
airline_graph = GraphFrame(g.vertices, airline_relationships)

clusters = airline_graph.labelPropagation(maxIter=10)
(clusters
 .sort("label")
 .groupby("label")
 .agg(F.collect_list("id").alias("airports"),
      F.count("id").alias("count"))
 .sort("count", ascending=False)
 .show(truncate=70, n=10))
```

If we run that query we'll see the following output:

label	airports	count
1606317768706	[IND, ORF, ATW, RIC, TRI, XNA, ECP, AVL, JAX, SYR, BHM, GSO, MEM, C...	89
1219770712067	[GEG, SLC, DTW, LAS, SEA, BOS, MSN, SNA, JFK, TVC, LIH, JAC, FLL, M...	53
17179869187	[RHV]	1
25769803777	[CWT]	1
25769803776	[CDW]	1
25769803782	[KNW]	1
25769803778	[DRT]	1
25769803779	[FOK]	1
25769803781	[HVR]	1
42949672962	[GTF]	1

Most of the airports DL uses have clustered into two groups; let's drill down into those. There are too many airports to show here, so we'll just show the airports with the biggest degree (ingoing and outgoing flights). We can write the following code to calculate airport degree:

```
all_flights = g.degrees.withColumnRenamed("id", "aId")
```

We'll then combine this with the airports that belong to the largest cluster:

```
(clusters
  .filter("label=1606317768706")
  .join(all_flights, all_flights.aId == clusters.id)
  .sort("degree", ascending=False)
  .select("id", "name", "degree")
  .show(truncate=False))
```

If we run that query we'll get this output:

id	name	degree
DFW	Dallas Fort Worth International Airport	47514
CLT	Charlotte Douglas International Airport	40495
IAH	George Bush Intercontinental Houston Airport	28814
EWR	Newark Liberty International Airport	25131
PHL	Philadelphia International Airport	20804
BWI	Baltimore/Washington International Thurgood Marshall Airport	18989
MDW	Chicago Midway International Airport	15178
BNA	Nashville International Airport	12455
DAL	Dallas Love Field	12084
IAD	Washington Dulles International Airport	11566

id	name	degree
STL	Lambert St Louis International Airport	11439
HOU	William P Hobby Airport	9742
IND	Indianapolis International Airport	8543
PIT	Pittsburgh International Airport	8410
CLE	Cleveland Hopkins International Airport	8238
CMH	Port Columbus International Airport	7640
SAT	San Antonio International Airport	6532
JAX	Jacksonville International Airport	5495
BDL	Bradley International Airport	4866
RSW	Southwest Florida International Airport	4569

In Figure 7-15 we can see that this cluster is actually focused on the East Coast to the Midwest of the United States.

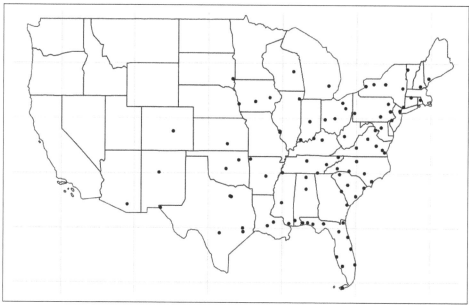

Figure 7-15. Cluster 1606317768706 airports

And now let's do the same thing with the second-largest cluster:

```
(clusters
 .filter("label=1219770712067")
 .join(all_flights, all_flights.aId == result.id)
 .sort("degree", ascending=False)
 .select("id", "name", "degree")
 .show(truncate=False))
```

If we run that query we get this output:

id	name	degree
ATL	Hartsfield Jackson Atlanta International Airport	67672
ORD	Chicago O'Hare International Airport	56681
DEN	Denver International Airport	39671
LAX	Los Angeles International Airport	38116
PHX	Phoenix Sky Harbor International Airport	30206
SFO	San Francisco International Airport	29865
LGA	La Guardia Airport	29416
LAS	McCarran International Airport	27801
DTW	Detroit Metropolitan Wayne County Airport	27477
MSP	Minneapolis-St Paul International/Wold-Chamberlain Airport	27163
BOS	General Edward Lawrence Logan International Airport	26214
SEA	Seattle Tacoma International Airport	24098
MCO	Orlando International Airport	23442
JFK	John F Kennedy International Airport	22294
DCA	Ronald Reagan Washington National Airport	22244
SLC	Salt Lake City International Airport	18661
FLL	Fort Lauderdale Hollywood International Airport	16364
SAN	San Diego International Airport	15401
MIA	Miami International Airport	14869
TPA	Tampa International Airport	12509

In Figure 7-16 we can see that this cluster is apparently more hub-focused, with some additional northwestern stops along the way.

Figure 7-16. Cluster 1219770712067 airports

The code we used to generate these maps is available in the book's GitHub repository (*https://bit.ly/2FPgGVV*).

When checking the DL website for frequent flyer programs, we notice a use-two-get-one-free promotion. If we use our points for two flights, we get another for free—but only if we fly within one of the two clusters! Perhaps it's a better use of our time, and certainly our points, to stay in a cluster.

Reader Exercises

- Use a Shortest Path algorithm to evaluate the number of flights from your home to the Bozeman Yellowstone International Airport (BZN).
- Are there any differences if you use relationship weights?

Summary

In the last few chapters we've provided details on how key graph algorithms for path-finding, centrality, and community detection work in Apache Spark and Neo4j. In this chapter we walked through workflows that included using several algorithms in context with other tasks and analysis. We used a travel business scenario to analyze Yelp data in Neo4j and a personal air travel scenario to evaluate US airline data in Spark.

Next, we'll look at a use for graph algorithms that's becoming increasingly important: graph-enhanced machine learning.

CHAPTER 8

Using Graph Algorithms to Enhance Machine Learning

We've covered several algorithms that learn and update state at each iteration, such as Label Propagation; however, up until this point, we've emphasized graph algorithms for general analytics. Because there's increasing application of graphs in machine learning (ML), we'll now look at how graph algorithms can be used to enhance ML workflows.

In this chapter, we focus on the most practical way to start improving ML predictions using graph algorithms: connected feature extraction and its use in predicting relationships. First, we'll cover some basic ML concepts and the importance of contextual data for better predictions. Then there's a quick survey of ways graph features are applied, including uses for spammer fraud, detection, and link prediction.

We'll demonstrate how to create a machine learning pipeline and then train and evaluate a model for link prediction, integrating Neo4j and Spark in our workflow. Our example will be based on the Citation Network Dataset, which contains authors, papers, author relationships, and citation relationships. We'll use several models to predict whether research authors are likely to collaborate in the future, and show how graph algorithms improve the results.

Machine Learning and the Importance of Context

Machine learning is not artificial intelligence (AI), but a method for achieving AI. ML uses algorithms to train software through specific examples and progressive improvements based on expected outcome—without explicit programming of how to accomplish these better results. Training involves providing a lot of data to a model and enabling it to learn how to process and incorporate that information.

In this sense, learning means that algorithms iterate, continually making changes to get closer to an objective goal, such as reducing classification errors in comparison to the training data. ML is also dynamic, with the ability to modify and optimize itself when presented with more data. This can take place in pre-usage training on many batches or as online learning during usage.

Recent successes in ML predictions, accessibility of large datasets, and parallel compute power have made ML more practical for those developing probabilistic models for AI applications. As machine learning becomes more widespread, it's important to remember its fundamental goal: making choices similarly to the way humans do. If we forget that goal, we may end up with just another version of highly targeted, rules-based software.

In order to increase machine learning accuracy while also making solutions more broadly applicable, we need to incorporate a lot of contextual information—just as people should use context for better decisions. Humans use their surrounding context, not just direct data points, to figure out what's essential in a situation, estimate missing information, and determine how to apply lessons to new situations. Context helps us improve predictions.

Graphs, Context, and Accuracy

Without peripheral and related information, solutions that attempt to predict behavior or make recommendations for varying circumstances require more exhaustive training and prescriptive rules. This is partly why AI is good at specific, well-defined tasks, but struggles with ambiguity. Graph-enhanced ML can help fill in that missing contextual information that is so important for better decisions.

We know from graph theory and from real life that relationships are often the strongest predictors of behavior. For example, if one person votes, there's an increased likelihood that their friends, family, and even coworkers will vote. Figure 8-1 illustrates a ripple effect based on reported voting and Facebook friends from the 2012 research paper "A 61-Million-Person Experiment in Social Influence and Political Mobilization" (*https://www.nature.com/articles/nature11421*), by R. Bond et al.

The authors found that friends reporting voting influenced an additional 1.4% of users to also claim they'd voted and, interestingly, friends of friends added another 1.7%. Small percentages can have a significant impact, and we can see in Figure 8-1 that people at two hops out had in total more impact than the direct friends alone. Voting and other examples of how our social networks impact us are covered in the book *Connected*, by Nicholas Christakis and James Fowler (Little, Brown and Company).

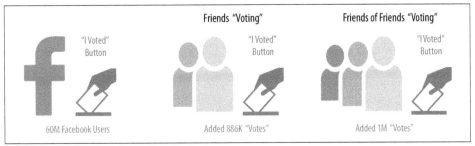

Figure 8-1. People are influenced to vote by their social networks. In this example, friends two hops away had more total impact than direct relationships.

Adding graph features and context improves predictions, especially in situations where connections matter. For example, retail companies personalize product recommendations with not only historical data but also contextual data about customer similarities and online behavior. Amazon's Alexa uses several layers of contextual models (*https://amzn.to/2YmSvqn*) that demonstrate improved accuracy. In 2018, Amazon also introduced "context carryover" to incorporate previous references in a conversation when answering new questions.

Unfortunately, many machine learning approaches today miss a lot of rich contextual information. This stems from ML's reliance on input data built from tuples, leaving out a lot of predictive relationships and network data. Furthermore, contextual information is not always readily available or is too difficult to access and process. Even finding connections that are four or more hops away can be a challenge at scale for traditional methods. Using graphs, we can more easily reach and incorporate connected data.

Connected Feature Engineering

Connected (Graph) Feature Engineering helps us take raw data and create a suitable subset and format for training our machine learning models. It's a foundational step that, when well executed, leads to ML that produces more consistently accurate predictions. Graph feature engineering includes feature extraction and selection.

Feature Extraction and Selection

Feature extraction is a way to distill large volumes of data and attributes down to a set of representative descriptive attributes. The process derives numerical values (features) for distinctive characteristics or patterns in input data so that we can differentiate categories in other data. It's used when data is difficult for a model to analyze directly—perhaps because of size, format, or the need for incidental comparisons.

Feature selection is the process of determining the subset of extracted features that are most important or influential to a target goal. It's used to surface predictive importance as well as for efficiency. For example, if we have 20 features and 13 of them together explain 92% of what we need to predict, we can eliminate 7 features in our model.

Putting together the right mix of features can increase accuracy because it fundamentally influences how our models learn. Because even modest improvements can make a significant difference, our focus in this chapter is on *connected features*. Connected features are features extracted from the structure of the data. These features can be derived from graph-local queries based on parts of the graph surrounding a node, or graph-global queries that use graph algorithms to identify predictive elements within data based on relationships for connected feature extraction.

And it's not only important to get the right combination of features, but also to eliminate unnecessary features to reduce the likelihood that our models will be hypertargeted. This keeps us from creating models that only work well on our training data (known as *overfitting*) and significantly expands applicability. We can also use graph algorithms to evaluate those features and determine which ones are most influential to our model for connected feature selection. For example, we can map features to nodes in a graph, create relationships based on similar features, and then compute the centrality of features. Feature relationships can be defined by the ability to preserve cluster densities of data points. This method is described using datasets with high dimension and low sample size in "Unsupervised Graph-Based Feature Selection Via Subspace and PageRank Centrality" (*https://bit.ly/2HGON5B*), by K. Henniab, N. Mezghani, and C. Gouin-Vallerand.

Graph Embedding

Graph embedding is the representation of the nodes and relationships in a graph as *feature vectors*. These are merely collections of features that have dimensional mappings, such as the (x,y,z) coordinates shown in Figure 8-2.

Graph Representation   Matrix Representation   Vector Representation
                                                                       n -dimensional vector space

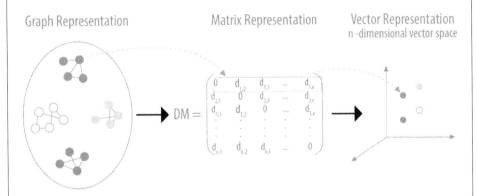

$$DM = \begin{pmatrix} 0 & d_{1,2} & d_{1,3} & \cdots & d_{1,n} \\ d_{2,1} & 0 & d_{2,3} & \cdots & d_{2,n} \\ d_{3,1} & d_{3,2} & 0 & \cdots & d_{3,n} \\ \vdots & \vdots & \vdots & \vdots & \vdots \\ d_{n,1} & d_{n,2} & d_{n,3} & \cdots & 0 \end{pmatrix}$$

Figure 8-2. Graph embedding maps graph data into feature vectors that can be visualized as multidimensional coordinates.

Graph embedding uses graph data slightly differently than in connected feature extraction. It enables us to represent entire graphs, or subsets of graph data, in a numerical format ready for machine learning tasks. This is especially useful for unsupervised learning, where the data is not categorized because it pulls in more contextual information through relationships. Graph embedding is also useful for data exploration, computing similarity between entities, and reducing dimensionality to aid in statistical analysis.

This is a quickly evolving space with several options, including node2vec, struc2vec, GraphSAGE (*https://bit.ly/2HYdhqH*), DeepWalk (*https://bit.ly/2JDmIOo*), and DeepGL (*https://bit.ly/2OryHxg*).

Now let's look at some of the types of connected features and how they are used.

Graphy Features

Graphy features include any number of connection-related metrics about our graph, such as the number of relationships going into or out of nodes, a count of potential triangles, and neighbors in common. In our example, we'll start with these measures because they are simple to gather and a good test of early hypotheses.

In addition, when we know precisely what we're looking for, we can use feature engineering. For instance, if we want to know how many people have a fraudulent account at up to four hops out. This approach uses graph traversal to very efficiently find deep paths of relationships, looking at things such as labels, attributes, counts, and inferred relationships.

We can also easily automate these processes and deliver those predictive graphy features into our existing pipeline. For example, we could abstract a count of fraudster relationships and add that number as a node attribute to be used for other machine learning tasks.

Graph Algorithm Features

We can also use graph algorithms to find features where we know the general structure we're looking for but not the exact pattern. As an illustration, let's say we know certain types of community groupings are indicative of fraud; perhaps there's a prototypical density or hierarchy of relationships. In this case, we don't want a rigid feature of an exact organization but rather a flexible and globally relevant structure. We'll use community detection algorithms to extract connected features in our example, but centrality algorithms, like PageRank, are also frequently applied.

Furthermore, approaches that combine several types of connected features seem to outperform sticking to one single method. For example, we could combine connected features to predict fraud with indicators based on communities found via the Louvain algorithm, influential nodes using PageRank, and the measure of known fraudsters at three hops out.

A combined approach is shown in Figure 8-3; the authors combine graph algorithms like PageRank and Coloring with graphy measure such as in-degree and out-degree. This diagram is taken from the paper "Collective Spammer Detection in Evolving Multi-Relational Social Networks" (*https://bit.ly/2TyG6Mm*), by S. Fakhraei et al.

The Graph Structure section illustrates connected feature extraction using several graph algorithms. Interestingly, the authors found extracting connected features from multiple types of relationships even more predictive than simply adding more features. The Report Subgraph section shows how graph features are converted into features that the ML model can use. By combining multiple methods in a graph-enhanced ML workflow, the authors were able to improve prior detection methods and classify 70% of spammers that had previously required manual labeling, with 90% accuracy.

Even once we have extracted connected features, we can improve our training by using graph algorithms like PageRank to prioritize the features with the most influence. This enables us to adequately represent our data while eliminating noisy variables that could degrade results or slow processing. With this type of information, we can also identify features with high co-occurrence for further model tuning via feature reduction. This method is outlined in the research paper "Using PageRank in Feature Selection" (*https://bit.ly/2JDDwVw*), by D. Ienco, R. Meo, and M. Botta.

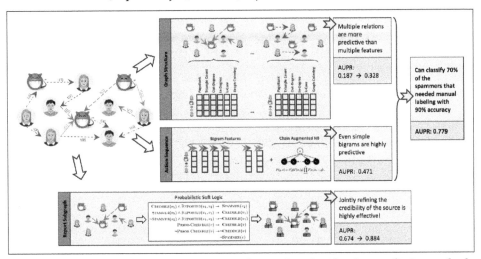

Figure 8-3. Connected feature extraction can be combined with other predictive methods to improve results. AUPR refers to the area under the precision-recall curve, with higher numbers preferred.

We've discussed how connected features are applied to scenarios involving fraud and spammer detection. In these situations, activities are often hidden in multiple layers of obfuscation and network relationships. Traditional feature extraction and selection methods may be unable to detect that behavior without the contextual information that graphs bring.

Another area where connected features enhance machine learning (and the focus of the rest of this chapter) is *link prediction*. Link prediction is a way to estimate how likely a relationship is to form in the future, or whether it should already be in our graph but is missing due to incomplete data. Since networks are dynamic and can grow fairly quickly, being able to predict links that will soon be added has broad applicability, from product recommendations to drug retargeting and even inferring criminal relationships.

Connected features from graphs are often used to improve link prediction using basic graphy features as well as features extracted from centrality and community algorithms. Link prediction based on node proximity or similarity is also standard; in the paper "The Link Prediction Problem for Social Networks" (*https://bit.ly/2uoyB0q*) D. Liben-Nowell and J. Kleinberg suggest that the network structure alone may contain enough latent information to detect node proximity and outperform more direct measures.

Now that we've looked at ways connected features can enhance machine learning, let's dive into our link prediction example and see how we can apply graph algorithms to improve our predictions.

Graphs and Machine Learning in Practice: Link Prediction

The rest of the chapter will demonstrate a hands-on example, based on the Citation Network Dataset (*https://aminer.org/citation*), a research dataset extracted from DBLP, ACM, and MAG. The dataset is described in the paper "ArnetMiner: Extraction and Mining of Academic Social Networks" (*http://bit.ly/2U4C3fb*), by J. Tang et al. The latest version contains 3,079,007 papers, 1,766,547 authors, 9,437,718 author relationships, and 25,166,994 citation relationships.

We'll work with a subset focused on articles from the following publications:

- *Lecture Notes in Computer Science*
- *Communications of the ACM*
- *International Conference on Software Engineering*
- *Advances in Computing and Communications*

Our resulting dataset contains 51,956 papers, 80,299 authors, 140,575 author relationships, and 28,706 citation relationships. We'll create a coauthors graph based on authors who have collaborated on papers and then predict future collaborations between pairs of authors. We're only interested in collaborations between authors who haven't collaborated before—we're not concerned with multiple collaborations between pairs of authors.

In the rest of the chapter, we'll set up the required tools and import the data into Neo4j. Then we'll cover how to properly balance data and split samples into Spark DataFrames for training and testing. After that, we explain our hypothesis and methods for link prediction before creating a machine learning pipeline in Spark. Finally, we'll walk through training and evaluating various prediction models, starting with basic graphy features and adding graph algorithm features extracted using Neo4j.

Tools and Data

Let's get started by setting up our tools and data. Then we'll explore our dataset and create a machine learning pipeline.

Before we do anything else, let's set up the libraries used in this chapter:

py2neo
  A Neo4j Python library that integrates well with the Python data science ecosystem

pandas
  A high-performance library for data wrangling outside of a database with easy-to-use data structures and data analysis tools

Spark MLlib
  Spark's machine learning library

 We use MLlib as an example of a machine learning library. The approach shown in this chapter could be used in combination with other ML libraries, such as scikit-learn.

All the code shown will be run within the pyspark REPL. We can launch the REPL by running the following command:

```
export SPARK_VERSION="spark-2.4.0-bin-hadoop2.7"
./${SPARK_VERSION}/bin/pyspark \
  --driver-memory 2g \
  --executor-memory 6g \
  --packages julioasotodv:spark-tree-plotting:0.2
```

This is similar to the command we used to launch the REPL in Chapter 3, but instead of GraphFrames, we're loading the `spark-tree-plotting` package. At the time of writing the latest released version of Spark is *spark-2.4.0-bin-hadoop2.7*, but as that may have changed by the time you read this, be sure to change the SPARK_VERSION environment variable appropriately.

Once we've launched that we'll import the following libraries that we'll be using:

```
from py2neo import Graph
import pandas as pd
from numpy.random import randint

from pyspark.ml import Pipeline
from pyspark.ml.classification import RandomForestClassifier
from pyspark.ml.feature import StringIndexer, VectorAssembler
from pyspark.ml.evaluation import BinaryClassificationEvaluator

from pyspark.sql.types import *
from pyspark.sql import functions as F

from sklearn.metrics import roc_curve, auc
from collections import Counter
```

```
from cycler import cycler
import matplotlib
matplotlib.use('TkAgg')
import matplotlib.pyplot as plt
```

And now let's create a connection to our Neo4j database:

```
graph = Graph("bolt://localhost:7687", auth=("neo4j", "neo"))
```

Importing the Data into Neo4j

Now we're ready to load the data into Neo4j and create a balanced split for our train-ing and testing. We need to download the ZIP file of Version 10 (*https://bit.ly/2TszAH3*) of the dataset, unzip it, and place the contents in our *import* folder. We should have the following files:

- *dblp-ref-0.json*
- *dblp-ref-1.json*
- *dblp-ref-2.json*
- *dblp-ref-3.json*

Once we have those files in the *import* folder, we need to add the following property to our Neo4j settings file so that we can process them using the APOC library:

```
apoc.import.file.enabled=true
apoc.import.file.use_neo4j_config=true
```

First we'll create constraints to ensure we don't create duplicate articles or authors:

```
CREATE CONSTRAINT ON (article:Article)
ASSERT article.index IS UNIQUE;

CREATE CONSTRAINT ON (author:Author)
ASSERT author.name IS UNIQUE;
```

Now we can run the following query to import the data from the JSON files:

```
CALL apoc.periodic.iterate(
  'UNWIND ["dblp-ref-0.json","dblp-ref-1.json",
          "dblp-ref-2.json","dblp-ref-3.json"] AS file
  CALL apoc.load.json("file:///" + file)
  YIELD value
  WHERE value.venue IN ["Lecture Notes in Computer Science",
                        "Communications of The ACM",
                        "international conference on software engineering",
                        "advances in computing and communications"]
  return value',
  'MERGE (a:Article {index:value.id})
  ON CREATE SET a += apoc.map.clean(value,["id","authors","references"],[0])
  WITH a,value.authors as authors
  UNWIND authors as author
```

```
    MERGE (b:Author{name:author})
    MERGE (b)<-[:AUTHOR]-(a)'
, {batchSize: 10000, iterateList: true});
```

This results in the graph schema seen in Figure 8-4.

Figure 8-4. The citation graph

This is a simple graph that connects articles and authors, so we'll add more informa-tion we can infer from relationships to help with predictions.

The Coauthorship Graph

We want to predict future collaborations between authors, so we'll start by creating a coauthorship graph. The following Neo4j Cypher query will create a CO_AUTHOR rela-tionship between every pair of authors that have collaborated on a paper:

```
MATCH (a1)<-[:AUTHOR]-(paper)-[:AUTHOR]->(a2:Author)
WITH a1, a2, paper
ORDER BY a1, paper.year
WITH a1, a2, collect(paper)[0].year AS year, count(*) AS collaborations
MERGE (a1)-[coauthor:CO_AUTHOR {year: year}]-(a2)
SET coauthor.collaborations = collaborations;
```

The year property that we set on the CO_AUTHOR relationship in the query is the earli-est year when those two authors collaborated. We're only interested in the first time that a pair of authors have collaborated—subsequent collaborations aren't relevant.

Figure 8-5 is in an example of part of the graph that gets created. We can already see some interesting community structures.

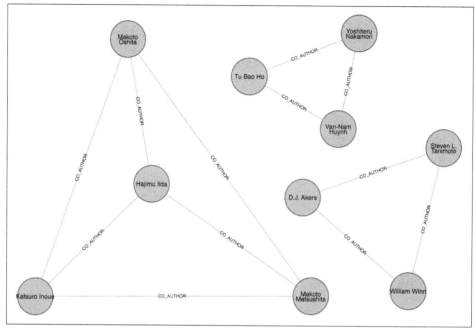

Figure 8-5. The coauthor graph

Each circle in this diagram represents one author and the lines between them are CO_AUTHOR relationships, so we have four authors that have all collaborated with each other on the left, and then on the right two examples of three authors who have collaborated. Now that we have our data loaded and a basic graph, let's create the two datasets we'll need for training and testing.

Creating Balanced Training and Testing Datasets

With link prediction problems we want to try and predict the future creation of links. This dataset works well for that because we have dates on the articles that we can use to split our data. We need to work out which year we'll use to define our training/test split. We'll train our model on everything before that year and then test it on the links created after that date.

Let's start by finding out when the articles were published. We can write the following query to get a count of the number of articles, grouped by year:

```
query = """
MATCH (article:Article)
RETURN article.year AS year, count(*) AS count
ORDER BY year
"""

by_year = graph.run(query).to_data_frame()
```

Let's visualize this as a bar chart, with the following code:

```
plt.style.use('fivethirtyeight')
ax = by_year.plot(kind='bar', x='year', y='count', legend=None, figsize=(15,8))
ax.xaxis.set_label_text("")
plt.tight_layout()
plt.show()
```

We can see the chart generated by executing this code in Figure 8-6.

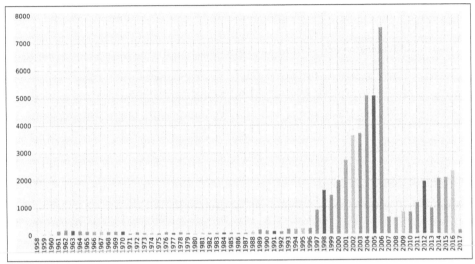

Figure 8-6. Articles by year

Very few articles were published before 1997, and then there were a lot published between 2001 and 2006, before a dip and then a gradual climb since 2011 (excluding 2013). It looks like 2006 could be a good year to split our data for training our model and making predictions. Let's check how many papers were published before that year and how many during and after. We can write the following query to compute this:

```
MATCH (article:Article)
RETURN article.year < 2006 AS training, count(*) AS count
```

The result of this is as follows, where *true* means a paper was published before 2006:

training	count
false	21059
true	30897

Not bad! 60% of the papers were published before 2006 and 40% during or after 2006. This is a fairly balanced split of data for our training and testing.

So now that we have a good split of papers, let's use the same 2006 split for coauthorship. We'll create a CO_AUTHOR_EARLY relationship between pairs of authors whose first collaboration was *before 2006*:

```
MATCH (a1)-[coAuthor:CO_AUTHOR]-(a2:Author)
WHERE coAuthor.year < 2006
MERGE (a1)-[coauthorEarly:CO_AUTHOR_EARLY {year: coAuthor.year}]-(a2)
SET coauthorEarly.collaborations = coAuthor.collaborations;
```

And then we'll create a CO_AUTHOR_LATE relationship between pairs of authors whose first collaboration was *during or after 2006*:

```
MATCH (a1)-[coAuthor:CO_AUTHOR]-(a2:Author)
WHERE coAuthor.year >= 2006
MERGE (a1)-[coauthorLate:CO_AUTHOR_LATE {year: coAuthor.year}]-(a2)
SET coauthorLate.collaborations = coAuthor.collaborations;
```

Before we build our training and test sets, let's check how many pairs of nodes we have that have links between them. The following query will find the number of CO_AUTHOR_EARLY pairs:

```
MATCH ()-[:CO_AUTHOR_EARLY]->()
RETURN count(*) AS count
```

Running that query will return the result shown here:

count

81096

And this query will find the number of CO_AUTHOR_LATE pairs:

```
MATCH ()-[:CO_AUTHOR_LATE]->()
RETURN count(*) AS count
```

Running that query returns this result:

count

74128

Now we're ready to build our training and test datasets.

Balancing and splitting data

The pairs of nodes with CO_AUTHOR_EARLY and CO_AUTHOR_LATE relationships between them will act as our positive examples, but we'll also need to create some negative examples. Most real-world networks are sparse, with concentrations of relationships, and this graph is no different. The number of examples where two nodes do not have a relationship is much larger than the number that do have a relationship.

If we query our CO_AUTHOR_EARLY data, we'll find there are 45,018 authors with that type of relationship but only 81,096 relationships between authors. That might not sound imbalanced, but it is: the potential maximum number of relationships that our graph could have is (45018 * 45017) / 2 = 1,013,287,653, which means there are a lot of negative examples (no links). If we used all the negative examples to train our model, we'd have a severe class imbalance problem. A model could achieve extremely high accuracy by predicting that every pair of nodes doesn't have a relationship.

In their paper "New Perspectives and Methods in Link Prediction" (*https://ntrda.me/ 2TrSg9K*), R. Lichtenwalter, J. Lussier, and N. Chawla describe several methods to address this challenge. One of these approaches is to build negative examples by finding nodes within our neighborhood that we aren't currently connected to.

We will build our negative examples by finding pairs of nodes that are a mix of between two and three hops away from each other, excluding those pairs that already have a relationship. We'll then downsample those pairs of nodes so that we have an equal number of positive and negative examples.

We have 314,248 pairs of nodes that don't have a relationship between each other at a distance of two hops. If we increase the distance to three hops, we have 967,677 pairs of nodes.

The following function will be used to downsample the negative examples:

```
def down_sample(df):
    copy = df.copy()
    zero = Counter(copy.label.values)[0]
    un = Counter(copy.label.values)[1]
    n = zero - un
    copy = copy.drop(copy[copy.label == 0].sample(n=n, random_state=1).index)
    return copy.sample(frac=1)
```

This function works out the difference between the number of positive and negative examples, and then samples the negative examples so that there are equal numbers. We can then run the following code to build a training set with balanced positive and negative examples:

```
train_existing_links = graph.run("""
MATCH (author:Author)-[:CO_AUTHOR_EARLY]->(other:Author)
RETURN id(author) AS node1, id(other) AS node2, 1 AS label
""").to_data_frame()

train_missing_links = graph.run("""
MATCH (author:Author)
WHERE (author)-[:CO_AUTHOR_EARLY]-()
MATCH (author)-[:CO_AUTHOR_EARLY*2..3]-(other)
WHERE not((author)-[:CO_AUTHOR_EARLY]-(other))
```

```
RETURN id(author) AS node1, id(other) AS node2, 0 AS label
""").to_data_frame()

train_missing_links = train_missing_links.drop_duplicates()
training_df = train_missing_links.append(train_existing_links, ignore_index=True)
training_df['label'] = training_df['label'].astype('category')
training_df = down_sample(training_df)
training_data = spark.createDataFrame(training_df)
```

We've now coerced the label column to be a category, where 1 indicates that there is a link between a pair of nodes, and 0 indicates that there is not a link. We can look at the data in our DataFrame by running the following code:

```
training_data.show(n=5)
```

node1	node2	label
10019	28091	1
10170	51476	1
10259	17140	0
10259	26047	1
10293	71349	1

The results show us a list of node pairs and whether they have a coauthor relationship; for example, nodes 10019 and 28091 have a 1 label, indicating a collaboration.

Now let's execute the following code to check the summary of contents for the Data-Frame:

```
training_data.groupby("label").count().show()
```

Here's the result:

label	count
0	81096
1	81096

We've created our training set with the same number of positive and negative samples. Now we need to do the same for the test set. The following code will build a test set with balanced positive and negative examples:

```
test_existing_links = graph.run("""
MATCH (author:Author)-[:CO_AUTHOR_LATE]->(other:Author)
RETURN id(author) AS node1, id(other) AS node2, 1 AS label
""").to_data_frame()

test_missing_links = graph.run("""
MATCH (author:Author)
WHERE (author)-[:CO_AUTHOR_LATE]-()
```

```
MATCH (author)-[:CO_AUTHOR*2..3]-(other)
WHERE not((author)-[:CO_AUTHOR]-(other))
RETURN id(author) AS node1, id(other) AS node2, 0 AS label
""").to_data_frame()

test_missing_links = test_missing_links.drop_duplicates()
test_df = test_missing_links.append(test_existing_links, ignore_index=True)
test_df['label'] = test_df['label'].astype('category')
test_df = down_sample(test_df)
test_data = spark.createDataFrame(test_df)
```

We can execute the following code to check the contents of the DataFrame:

```
test_data.groupby("label").count().show()
```

Which gives the following result:

label	count
0	74128
1	74128

Now that we have balanced training and test datasets, let's look at our methods for predicting links.

How We Predict Missing Links

We need to start with some basic assumptions about what elements in our data might predict whether two authors will become coauthors at a later date. Our hypothesis would vary by domain and problem, but in this case, we believe the most predictive features will be related to communities. We'll begin with the assumption that the following elements increase the probability that authors become coauthors:

- More coauthors in common
- Potential triadic relationships between authors
- Authors with more relationships
- Authors in the same community
- Authors in the same, tighter community

We'll build graph features based on our assumptions and use those to train a binary classifier. *Binary classification* is a type of ML with the task of predicting which of two predefined groups an element belongs to based on a rule. We're using the classifier for the task of predicting whether a pair of authors will have a link or not, based on a classification rule. For our examples, a value of 1 means there is a link (coauthorship), and a value of 0 means there isn't a link (no coauthorship).

We'll implement our binary classifier as a random forest in Spark. A *random forest* is an ensemble learning method for classification, regression, and other tasks, as illustrated in Figure 8-7.

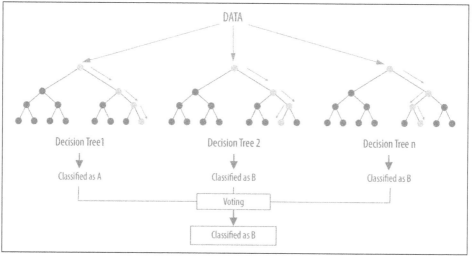

Figure 8-7. A random forest builds a collection of decision trees and then aggregates results for a majority vote (for classification) or an average value (for regression).

Our random forest classifier will take the results from the multiple decision trees we train and then use voting to predict a classification—in our example, whether there is a link (coauthorship) or not.

Now let's create our workflow.

Creating a Machine Learning Pipeline

We'll create our machine learning pipeline based on a random forest classifier in Spark. This method is well suited as our dataset will be comprised of a mix of strong and weak features. While the weak features will sometimes be helpful, the random forest method will ensure we don't create a model that only fits our training data.

To create our ML pipeline, we'll pass in a list of features as the fields variable—these are the features that our classifier will use. The classifier expects to receive those features as a single column called features, so we use the VectorAssembler to transform the data into the required format.

The following code creates a machine learning pipeline and sets up our parameters using MLlib:

```
def create_pipeline(fields):
    assembler = VectorAssembler(inputCols=fields, outputCol="features")
    rf = RandomForestClassifier(labelCol="label", featuresCol="features",
                                numTrees=30, maxDepth=10)
    return Pipeline(stages=[assembler, rf])
```

The `RandomForestClassifier` uses these parameters:

`labelCol`
> The name of the field containing the variable we want to predict; i.e., whether a pair of nodes have a link

`featuresCol`
> The name of the field containing the variables that will be used to predict whether a pair of nodes have a link

`numTrees`
> The number of decision trees that form the random forest

`maxDepth`
> The maximum depth of the decision trees

We chose the number of decision trees and their depth based on experimentation. We can think about hyperparameters like the settings of an algorithm that can be adjusted to optimize performance. The best hyperparameters are often difficult to determine ahead of time, and tuning a model usually requires some trial and error.

We've covered the basics and set up our pipeline, so let's dive into creating our model and evaluating how well it performs.

Predicting Links: Basic Graph Features

We'll start by creating a simple model that tries to predict whether two authors will have a future collaboration based on features extracted from common authors, preferential attachment, and the total union of neighbors:

Common authors
> Finds the number of potential triangles between two authors. This captures the idea that two authors who have coauthors in common may be introduced and collaborate in the future.

Preferential attachment
> Produces a score for each pair of authors by multiplying the number of coauthors each has. The intuition is that authors are more likely to collaborate with someone who already coauthors a lot of papers.

Total union of neighbors
> Finds the total number of coauthors that each author has, minus the duplicates.

In Neo4j, we can compute these values using Cypher queries. The following function will compute these measures for the training set:

```
def apply_graphy_training_features(data):
    query = """
    UNWIND $pairs AS pair
    MATCH (p1) WHERE id(p1) = pair.node1
    MATCH (p2) WHERE id(p2) = pair.node2
    RETURN pair.node1 AS node1,
           pair.node2 AS node2,
           gds.alpha.linkprediction.commonNeighbors(p1, p2, {
             relationshipQuery: "CO_AUTHOR_EARLY"}) AS commonAuthors,
           gds.alpha.linkprediction.preferentialAttachment(p1, p2, {
             relationshipQuery: "CO_AUTHOR_EARLY"}) AS prefAttachment,
           gds.alpha.linkprediction.totalNeighbors(p1, p2, {
             relationshipQuery: "CO_AUTHOR_EARLY"}) AS totalNeighbours
    """
    pairs = [{
        "node1": row["node1"],
        "node2": row["node2"]
    } for row in data.collect()]
    features = spark.createDataFrame(
        graph.run(query, {
            "pairs": pairs
        }).to_data_frame())
    return data.join(features, ["node1", "node2"])
```

And the following function will compute them for the test set:

```
def apply_graphy_test_features(data):
    query = """
    UNWIND $pairs AS pair
    MATCH (p1) WHERE id(p1) = pair.node1
    MATCH (p2) WHERE id(p2) = pair.node2
    RETURN pair.node1 AS node1,
           pair.node2 AS node2,
           gds.alpha.linkprediction.commonNeighbors(p1, p2, {
             relationshipQuery: "CO_AUTHOR"}) AS commonAuthors,
           gds.alpha.linkprediction.preferentialAttachment(p1, p2, {
             relationshipQuery: "CO_AUTHOR"}) AS prefAttachment,
           gds.alpha.linkprediction.totalNeighbors(p1, p2, {
             relationshipQuery: "CO_AUTHOR"}) AS totalNeighbours
    """
    pairs = [{
        "node1": row["node1"],
        "node2": row["node2"]
    } for row in data.collect()]
    features = spark.createDataFrame(
        graph.run(query, {
            "pairs": pairs
        }).to_data_frame())
    return data.join(features, ["node1", "node2"])
```

Both of these functions take in a DataFrame that contains pairs of nodes in the columns node1 and node2. We then build an array of maps containing these pairs and compute each of the measures for each pair of nodes.

The UNWIND clause is particularly useful in this chapter for taking a large collection of node pairs and returning all their features in one query.

We can apply these functions in Spark to our training and test DataFrames with the following code:

```
training_data = apply_graphy_training_features(training_data)
test_data = apply_graphy_test_features(test_data)
```

Let's explore the data in our training set. The following code will plot a histogram of the frequency of commonAuthors:

```
plt.style.use('fivethirtyeight')
fig, axs = plt.subplots(1, 2, figsize=(18, 7), sharey=True)
charts = [(1, "have collaborated"), (0, "haven't collaborated")]

for index, chart in enumerate(charts):
    label, title = chart
    filtered = training_data.filter(training_data["label"] == label)
    common_authors = filtered.toPandas()["commonAuthors"]
    histogram = common_authors.value_counts().sort_index()
    histogram /= float(histogram.sum())
    histogram.plot(kind="bar", x='Common Authors', color="darkblue",
                   ax=axs[index], title=f"Authors who {title} (label={label})")
    axs[index].xaxis.set_label_text("Common Authors")

plt.tight_layout()
plt.show()
```

We can see the chart generated in Figure 8-8.

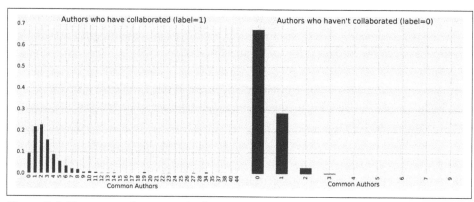

Figure 8-8. Frequency of commonAuthors

On the left we see the frequency of `commonAuthors` when authors have collaborated, and on the right we see the frequency of `commonAuthors` when they haven't. For those who haven't collaborated (right side) the maximum number of common authors is 9, but 95% of the values are 1 or 0. It's not surprising that of the people who have not collaborated on a paper, most also do not have many other coauthors in common. For those who have collaborated (left side), 70% have less than five coauthors in common, with a spike between one and two other coauthors.

Now we want to train a model to predict missing links. The following function does this:

```
def train_model(fields, training_data):
    pipeline = create_pipeline(fields)
    model = pipeline.fit(training_data)
    return model
```

We'll start by creating a basic model that only uses `commonAuthors`. We can create that model by running this code:

```
basic_model = train_model(["commonAuthors"], training_data)
```

With our model trained, let's check how it performs against some dummy data. The following code evaluates the code against different values for `commonAuthors`:

```
eval_df = spark.createDataFrame(
    [(0,), (1,), (2,), (10,), (100,)],
    ['commonAuthors'])

(basic_model.transform(eval_df)
 .select("commonAuthors", "probability", "prediction")
 .show(truncate=False))
```

Running that code will give the following result:

commonAuthors	probability	prediction
0	[0.7540494940434322,0.24595050595656787]	0.0
1	[0.7540494940434322,0.24595050595656787]	0.0
2	[0.0536835525078107,0.9463164474921892]	1.0
10	[0.0536835525078107,0.9463164474921892]	1.0

If we have a `commonAuthors` value of less than 2 there's a 75% probability that there won't be a relationship between the authors, so our model predicts 0. If we have a `commonAuthors` value of 2 or more there's a 94% probability that there will be a relationship between the authors, so our model predicts 1.

Let's now evaluate our model against the test set. Although there are several ways to evaluate how well a model performs, most are derived from a few baseline predictive metrics, as outlined in Table 8-1.

Table 8-1. Predictive metrics

Measure	Formula	Description
Accuracy	$\frac{TruePositives + TrueNegatives}{TotalPredictions}$	The fraction of predictions our model gets right, or the total number of correct predictions divided by the total number of predictions. Note that accuracy alone can be misleading, especially when our data is unbalanced. For example, if we have a dataset containing 95 cats and 5 dogs and our model predicts that every image is a cat we'll have a 95% accuracy score despite correctly identifying none of the dogs.
Precision	$\frac{TruePositives}{TruePositives + FalsePositives}$	The proportion of *positive identifications* that are correct. A low precision score indicates more false positives. A model that produces no false positives has a precision of 1.0.
Recall (true positive rate)	$\frac{TruePositives}{TruePositives + FalseNegatives}$	The proportion of *actual positives* that are identified correctly. A low recall score indicates more false negatives. A model that produces no false negatives has a recall of 1.0.
False positive rate	$\frac{FalsePositives}{FalsePositives + TrueNegatives}$	The proportion of *incorrect positives* that are identified. A high score indicates more false positives.
Receiver operating characteristic (ROC) curve	X-Y chart	ROC curve is a plot of the Recall (true positive rate) against the False Positive rate at different classification thresholds. The area under the curve (AUC) measures the two-dimensional area underneath the ROC curve from an X-Y axis (0,0) to (1,1).

We'll use accuracy, precision, recall, and ROC curves to evaluate our models. Accuracy is a coarse measure, so we'll focus on increasing our overall precision and recall measures. We'll use the ROC curves to compare how individual features change predictive rates.

Depending on our goals we may want to favor different measures. For example, we may want to eliminate all false negatives for disease indicators, but we wouldn't want to push predictions of everything into a positive result. There may be multiple thresholds we set for different models that pass some results through to secondary inspection on the likelihood of false results.

Lowering classification thresholds results in more overall positive results, thus increasing both false positives and true positives.

Let's use the following function to compute these predictive measures:

```
def evaluate_model(model, test_data):
    # Execute the model against the test set
    predictions = model.transform(test_data)

    # Compute true positive, false positive, false negative counts
    tp = predictions[(predictions.label == 1) &
                     (predictions.prediction == 1)].count()
    fp = predictions[(predictions.label == 0) &
                     (predictions.prediction == 1)].count()
    fn = predictions[(predictions.label == 1) &
                     (predictions.prediction == 0)].count()

    # Compute recall and precision manually
    recall = float(tp) / (tp + fn)
    precision = float(tp) / (tp + fp)

    # Compute accuracy using Spark MLLib's binary classification evaluator
    accuracy = BinaryClassificationEvaluator().evaluate(predictions)

    # Compute false positive rate and true positive rate using sklearn functions
    labels = [row["label"] for row in predictions.select("label").collect()]
    preds = [row["probability"][1] for row in predictions.select
                ("probability").collect()]
    fpr, tpr, threshold = roc_curve(labels, preds)
    roc_auc = auc(fpr, tpr)

    return { "fpr": fpr, "tpr": tpr, "roc_auc": roc_auc, "accuracy": accuracy,
            "recall": recall, "precision": precision }
```

We'll then write a function to display the results in an easier-to-consume format:

```
def display_results(results):
    results = {k: v for k, v in results.items() if k not in
                  ["fpr", "tpr", "roc_auc"]}
    return pd.DataFrame({"Measure": list(results.keys()),
                     "Score": list(results.values())})
```

We can call the function with this code and display the results:

```
basic_results = evaluate_model(basic_model, test_data)
display_results(basic_results)
```

The predictive measures for the common authors model are:

measure	score
accuracy	0.864457
recall	0.753278
precision	0.968670

This is not a bad start given that we're predicting future collaboration based only on the number of common authors in our pairs of authors. However, we get a bigger picture if we consider these measures in context with one another. For example, this model has a precision of 0.968670, which means it's very good at predicting that *links exist*. However, our recall is 0.753278, which means it's not good at predicting when *links do not exist*.

We can also plot the ROC curve (correlation of true positives and False positives) using the following functions:

```
def create_roc_plot():
    plt.style.use('classic')
    fig = plt.figure(figsize=(13, 8))
    plt.xlim([0, 1])
    plt.ylim([0, 1])
    plt.ylabel('True Positive Rate')
    plt.xlabel('False Positive Rate')
    plt.rc('axes', prop_cycle=(cycler('color',
                    ['r', 'g', 'b', 'c', 'm', 'y', 'k'])))
    plt.plot([0, 1], [0, 1], linestyle='--', label='Random score
                (AUC = 0.50)')
    return plt, fig


def add_curve(plt, title, fpr, tpr, roc):
    plt.plot(fpr, tpr, label=f"{title} (AUC = {roc:0.2})")
```

We call it like this:

```
plt, fig = create_roc_plot()

add_curve(plt, "Common Authors",
          basic_results["fpr"], basic_results["tpr"], basic_results["roc_auc"])

plt.legend(loc='lower right')
plt.show()
```

We can see the ROC curve for our basic model in Figure 8-9. The common authors model gives us a 0.86 area under the curve (AUC) score. Although this gives us one overall predictive measure, we need the chart (or other measures) to evaluate whether this fits our goal. In Figure 8-9 we see that as we get close to an 80% true positive rate (recall) our false positive rate reaches about 20%. That could be problematic in scenarios like fraud detection where false positives are expensive to chase.

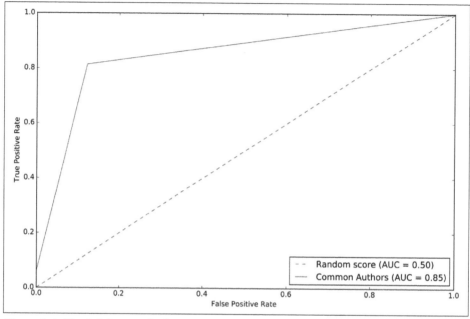

Figure 8-9. The ROC curve for basic model

Now let's use the other graphy features to see if we can improve our predictions. Before we train our model, let's see how the data is distributed. We can run the following code to show descriptive statistics for each of our graphy features:

```
(training_data.filter(training_data["label"]==1)
 .describe()
 .select("summary", "commonAuthors", "prefAttachment", "totalNeighbors")
 .show())

(training_data.filter(training_data["label"]==0)
 .describe()
 .select("summary", "commonAuthors", "prefAttachment", "totalNeighbors")
 .show())
```

We can see the results of running those bits of code in the following tables.

summary	commonAuthors	prefAttachment	totalNeighbors
count	81096	81096	81096
mean	3.5959233501035808	69.93537289138798	10.082408503502021
stddev	4.715942231635516	171.47092255919472	8.44109970920685
min	0	1	2
max	44	3150	90

summary	commonAuthors	prefAttachment	totalNeighbors
count	81096	81096	81096
mean	0.37666469369635985	48.18137762651672	12.97586810693499
stddev	0.6194576095461857	94.92635344980489	10.082991078685803
min	0	1	1
max	9	1849	89

Features with larger differences between links (coauthorship) and no link (no coauthorship) should be more predictive because the divide is greater. The average value for prefAttachment is higher for authors who have collaborated versus those who haven't. That difference is even more substantial for commonAuthors. We notice that there isn't much difference in the values for totalNeighbors, which probably means this feature won't be very predictive. Also interesting is the large standard deviation as well as the minimum and maximum values for preferential attachment. This is what we might expect for small-world networks with concentrated hubs (superconnectors).

Now let's train a new model, adding preferential attachment and total union of neighbors, by running the following code:

```
fields = ["commonAuthors", "prefAttachment", "totalNeighbors"]
graphy_model = train_model(fields, training_data)
```

And now let's evaluate the model and display the results:

```
graphy_results = evaluate_model(graphy_model, test_data)
display_results(graphy_results)
```

The predictive measures for the graphy model are:

measure	score
accuracy	0.978351
recall	0.924226
precision	0.943795

Our accuracy and recall have increased substantially, but the precision has dropped a bit and we're still misclassifying about 8% of the links. Let's plot the ROC curve and compare our basic and graphy models by running the following code:

```
plt, fig = create_roc_plot()

add_curve(plt, "Common Authors",
          basic_results["fpr"], basic_results["tpr"],
                            basic_results["roc_auc"])

add_curve(plt, "Graphy",
          graphy_results["fpr"], graphy_results["tpr"],
```

```
                              graphy_results["roc_auc"])

    plt.legend(loc='lower right')
    plt.show()
```

We can see the output in Figure 8-10.

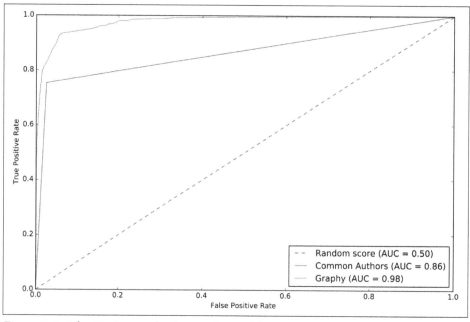

Figure 8-10. The ROC curve for the graphy model

Overall it looks like we're headed in the right direction and it's helpful to visualize comparisons to get a feel for how different models impact our results.

Now that we have more than one feature, we want to evaluate which features are making the most difference. We'll use *feature importance* to rank the impact of different features to our model's prediction. This enables us to evaluate the influence on results that different algorithms and statistics have.

 To compute feature importance, the random forest algorithm in Spark averages the reduction in impurity across all trees in the forest. The *impurity* is the frequency at which randomly assigned labels are incorrect.

Feature rankings are in comparison to the group of features we're evaluating, always normalized to 1. If we rank one feature, its feature importance is 1.0 as it has 100% of the influence on the model.

The following function creates a chart showing the most influential features:

```
def plot_feature_importance(fields, feature_importances):
    df = pd.DataFrame({"Feature": fields, "Importance": feature_importances})
    df = df.sort_values("Importance", ascending=False)
    ax = df.plot(kind='bar', x='Feature', y='Importance', legend=None)
    ax.xaxis.set_label_text("")
    plt.tight_layout()
    plt.show()
```

And we call it like this:

```
rf_model = graphy_model.stages[-1]
plot_feature_importance(fields, rf_model.featureImportances)
```

The results of running that function can be seen in Figure 8-11.

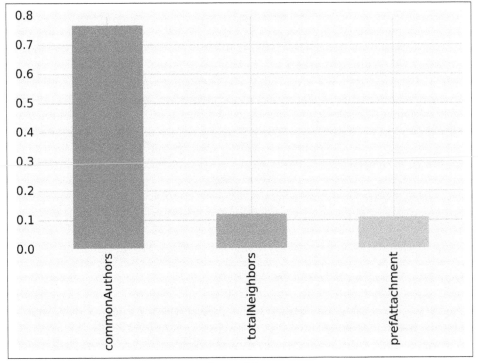

Figure 8-11. Feature importance: graphy model

Of the three features we've used so far, commonAuthors is the most important feature by a large margin.

To understand how our predictive models are created, we can visualize one of the decision trees in our random forest using the spark-tree-plotting library (*https://bit.ly/2usxOf2*). The following code generates a GraphViz file (*http://www.graphviz.org*):

```
from spark_tree_plotting import export_graphviz

dot_string = export_graphviz(rf_model.trees[0],
    featureNames=fields, categoryNames=[], classNames=["True", "False"],
    filled=True, roundedCorners=True, roundLeaves=True)

with open("/tmp/rf.dot", "w") as file:
    file.write(dot_string)
```

We can then generate a visual representation of that file by running the following command from the terminal:

```
dot -Tpdf /tmp/rf.dot -o /tmp/rf.pdf
```

The output of that command can be seen in Figure 8-12.

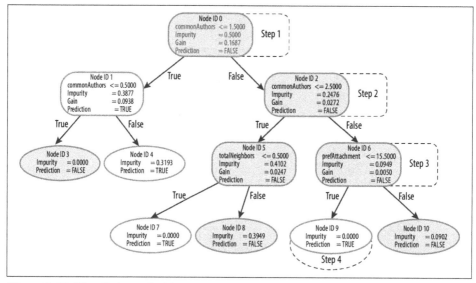

Figure 8-12. Visualizing a decision tree

Imagine that we're using this decision tree to predict whether a pair of nodes with the following features are linked:

commonAuthors	prefAttachment	totalNeighbors
10	12	5

Our random forest walks through several steps to create a prediction:

1. We start from node 0, where we have more than 1.5 commonAuthors, so we follow the False branch down to node 2.

2. We have more than 2.5 `commonAuthors` here, so we follow the `False` branch to node 6.

3. We have a score of less than 15.5 for `prefAttachment`, which takes us to node 9.

4. Node 9 is a leaf node in this decision tree, which means that we don't have to check any more conditions—the value of `Prediction` (i.e., `True`) on this node is the decision tree's prediction.

5. Finally, the random forest evaluates the item being predicted against a collection of these decision trees and makes its prediction based on the most popular outcome.

Now let's look at adding more graph features.

Predicting Links: Triangles and the Clustering Coefficient

Recommendation solutions often base predictions on some form of triangle metric, so let's see if they further help with our example. We can compute the number of triangles that a node is a part of by running the following query:

```
CALL gds.triangleCount.write({
  nodeProjection: 'Author',
  relationshipProjection: {
    CO_AUTHOR_EARLY: {
      type: 'CO_AUTHOR_EARLY',
      orientation: 'UNDIRECTED'
    }
  },
  writeProperty: 'trianglesTrain'
});

CALL gds.triangleCount.write({
  nodeProjection: 'Author',
  relationshipProjection: {
    CO_AUTHOR: {
      type: 'CO_AUTHOR',
      orientation: 'UNDIRECTED'
    }
  },
  writeProperty: 'trianglesTest'
});
```

And its clustering coefficient by running the following query:

```
CALL gds.localClusteringCoefficient.write({
  nodeProjection: 'Author',
  relationshipProjection: {
    CO_AUTHOR_EARLY: {
      type: 'CO_AUTHOR_EARLY',
      orientation: 'UNDIRECTED'
    }
  }
```

```
    },
    writeProperty: 'coefficientTrain'
});

CALL gds.localClusteringCoefficient.write({
    nodeProjection: 'Author',
    relationshipProjection: {
        CO_AUTHOR: {
            type: 'CO_AUTHOR',
            orientation: 'UNDIRECTED'
        }
    },
    writeProperty: 'coefficientTest'
});
```

The following function will add these features to our DataFrames:

```
def apply_triangles_features(data, triangles_prop, coefficient_prop):
    query = """
    UNWIND $pairs AS pair
    MATCH (p1) WHERE id(p1) = pair.node1
    MATCH (p2) WHERE id(p2) = pair.node2
    RETURN pair.node1 AS node1,
           pair.node2 AS node2,
           apoc.coll.min([p1[$trianglesProp], p2[$trianglesProp]])
                                        AS minTriangles,
           apoc.coll.max([p1[$trianglesProp], p2[$trianglesProp]])
                                        AS maxTriangles,
           apoc.coll.min([p1[$coefficientProp], p2[$coefficientProp]])
                                        AS minCoefficient,
           apoc.coll.max([p1[$coefficientProp], p2[$coefficientProp]])
                                        AS maxCoefficient
    """

    params = {
        "pairs": [{"node1": row["node1"], "node2": row["node2"]}
                             for row in data.collect()],
        "trianglesProp": triangles_prop,
        "coefficientProp": coefficient_prop
    }
    features = spark.createDataFrame(graph.run(query, params).to_data_frame())
    return data.join(features, ["node1", "node2"])
```

 Notice that we've used min and max prefixes for our triangle count
and clustering coefficient algorithms. We need a way to prevent our
model from learning based on the order authors in pairs are passed
in from our undirected graph. To do this, we've split these features
by the authors with minimum and maximum counts.

We can apply this function to our training and test DataFrames with the following
code:

```
training_data = apply_triangles_features(training_data,
                                "trianglesTrain", "coefficientTrain")
test_data = apply_triangles_features(test_data,
                                "trianglesTest", "coefficientTest")
```

And run this code to show descriptive statistics for each of our triangle features:

```
(training_data.filter(training_data["label"]==1)
  .describe()
  .select("summary", "minTriangles", "maxTriangles",
                  "minCoefficient", "maxCoefficient")
  .show())

(training_data.filter(training_data["label"]==0)
  .describe()
  .select("summary", "minTriangles", "maxTriangles", "minCoefficient",
                                    "maxCoefficient")
  .show())
```

We can see the results of running those bits of code in the following tables.

summary	minTriangles	maxTriangles	minCoefficient	maxCoefficient
count	81096	81096	81096	81096
mean	19.478260333431983	27.73590559337082	0.5703773654487051	0.8453786164620439
stddev	65.7615282768483	74.01896188921927	0.3614610553659958	0.2939681857356519
min	0	0	0.0	0.0
max	622	785	1.0	1.0

summary	minTriangles	maxTriangles	minCoefficient	maxCoefficient
count	81096	81096	81096	81096
mean	5.754661142349808	35.651980368945445	0.49048921333297446	0.860283935358397
stddev	20.639236521699	85.82843448272624	0.3684138346533951	0.2578219623967906
min	0	0	0.0	0.0
max	617	785	1.0	1.0

Notice in this comparison that there isn't as great a difference between the coauthorship and no-coauthorship data. This could mean that these features aren't as predictive. We can train another model by running the following code:

```
fields = ["commonAuthors", "prefAttachment", "totalNeighbors",
            "minTriangles", "maxTriangles", "minCoefficient", "maxCoefficient"]
triangle_model = train_model(fields, training_data)
```

And now let's evaluate the model and display the results:

```
triangle_results = evaluate_model(triangle_model, test_data)
display_results(triangle_results)
```

The predictive measures for the triangles model are shown in this table:

measure	score
accuracy	0.992924
recall	0.965384
precision	0.958582

Our predictive measures have increased well by adding each new feature to the previous model. Let's add our triangles model to our ROC curve chart with the following code:

```
plt, fig = create_roc_plot()

add_curve(plt, "Common Authors",
          basic_results["fpr"], basic_results["tpr"], basic_results["roc_auc"])

add_curve(plt, "Graphy",
          graphy_results["fpr"], graphy_results["tpr"],
                          graphy_results["roc_auc"])

add_curve(plt, "Triangles",
          triangle_results["fpr"], triangle_results["tpr"],
                          triangle_results["roc_auc"])

plt.legend(loc='lower right')
plt.show()
```

We can see the output in Figure 8-13.

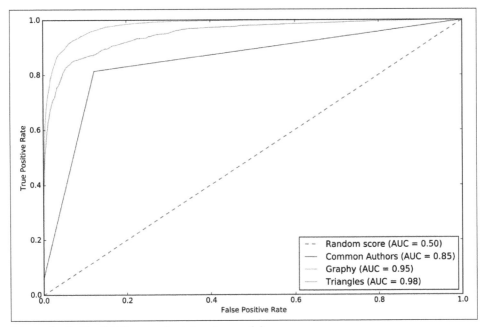

Figure 8-13. The ROC curve for triangles model

Our models have generally improved, and we're in the high 90s for predictive meas-ures. This is when things usually get difficult, because the easiest gains are made but there's still room for improvement. Let's see how the important features have changed:

```
rf_model = triangle_model.stages[-1]
plot_feature_importance(fields, rf_model.featureImportances)
```

The results of running that function can be seen in Figure 8-14. The common authors feature still has the greatest single impact on our model. Perhaps we need to look at new areas and see what happens when we add community information.

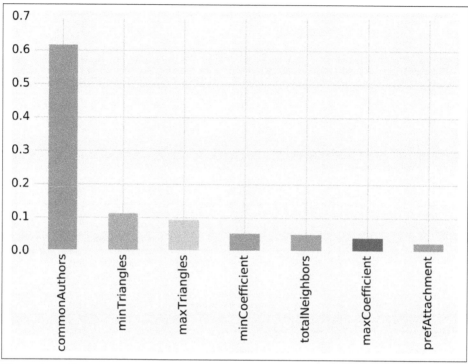

Figure 8-14. Feature importance: triangles model

Predicting Links: Community Detection

We hypothesize that nodes that are in the same community are more likely to have a link between them if they don't already. Moreover, we believe that the tighter a community is, the more likely links are.

First, we'll compute more coarse-grained communities using the Label Propagation algorithm in Neo4j. We do this by running the following query, which will store the community in the property `partitionTrain` for the training set and `partitionTest` for the test set:

```
CALL gds.labelPropagation.write({
  nodeProjection: "Author",
  relationshipProjection: {
    CO_AUTHOR_EARLY: {
      type: 'CO_AUTHOR_EARLY',
      orientation: 'UNDIRECTED'
    }
  },
  writeProperty: "partitionTrain"
});

CALL gds.labelPropagation.write({
```

```
    nodeProjection: "Author",
    relationshipProjection: {
      CO_AUTHOR: {
        type: 'CO_AUTHOR',
        orientation: 'UNDIRECTED'
      }
    },
    writeProperty: "partitionTest"
});
```

We'll also compute finer-grained groups using the Louvain algorithm. The Louvain algorithm returns intermediate clusters, and we'll store the smallest of these clusters in the property louvainTrain for the training set and louvainTest for the test set:

```
CALL gds.louvain.stream({
  nodeProjection: 'Author',
  relationshipProjection: {
    CO_AUTHOR_EARLY: {
      type: 'CO_AUTHOR_EARLY',
      orientation: 'UNDIRECTED'
    }
  },
  includeIntermediateCommunities: true
})
YIELD nodeId, communityId, intermediateCommunityIds
WITH gds.util.asNode(nodeId) AS node,
     intermediateCommunityIds[0] AS smallestCommunity
SET node.louvainTrain = smallestCommunity;

CALL gds.louvain.stream({
  nodeProjection: 'Author',
  relationshipProjection: {
    CO_AUTHOR: {
      type: 'CO_AUTHOR',
      orientation: 'UNDIRECTED'
    }
  },
  includeIntermediateCommunities: true
})
YIELD nodeId, communityId, intermediateCommunityIds
WITH gds.util.asNode(nodeId) AS node,
     intermediateCommunityIds[0] AS smallestCommunity
SET node.louvainTest = smallestCommunity;
```

We'll now create the following function to return the values from these algorithms:

```
def apply_community_features(data, partition_prop, louvain_prop):
    query = """
    UNWIND $pairs AS pair
    MATCH (p1) WHERE id(p1) = pair.node1
    MATCH (p2) WHERE id(p2) = pair.node2
    RETURN pair.node1 AS node1,
           pair.node2 AS node2,
```

```
                CASE WHEN p1[$partitionProp] = p2[$partitionProp] THEN
                        1 ELSE 0 END AS samePartition,
                CASE WHEN p1[$louvainProp] = p2[$louvainProp] THEN
                        1 ELSE 0 END AS sameLouvain
        """
    params = {
        "pairs": [{"node1": row["node1"], "node2": row["node2"]} for
                        row in data.collect()],
        "partitionProp": partition_prop,
        "louvainProp": louvain_prop
    }
    features = spark.createDataFrame(graph.run(query, params).to_data_frame())
    return data.join(features, ["node1", "node2"])
```

We can apply this function to our training and test DataFrames in Spark with the fol‐
lowing code:

```
training_data = apply_community_features(training_data,
                                "partitionTrain", "louvainTrain")
test_data = apply_community_features(test_data, "partitionTest", "louvainTest")
```

And we can run this code to see whether pairs of nodes belong in the same partition:

```
plt.style.use('fivethirtyeight')
fig, axs = plt.subplots(1, 2, figsize=(18, 7), sharey=True)
charts = [(1, "have collaborated"), (0, "haven't collaborated")]

for index, chart in enumerate(charts):
    label, title = chart
    filtered = training_data.filter(training_data["label"] == label)
    values = (filtered.withColumn('samePartition',
            F.when(F.col("samePartition") == 0, "False")
                            .otherwise("True"))
          .groupby("samePartition")
          .agg(F.count("label").alias("count"))
          .select("samePartition", "count")
          .toPandas())
    values.set_index("samePartition", drop=True, inplace=True)
    values.plot(kind="bar", ax=axs[index], legend=None,
                title=f"Authors who {title} (label={label})")
    axs[index].xaxis.set_label_text("Same Partition")

plt.tight_layout()
plt.show()
```

We see the results of running that code in Figure 8-15.

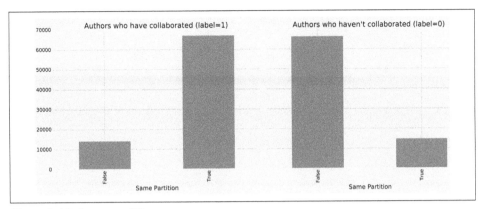

Figure 8-15. Same partitions

It looks like this feature could be quite predictive—authors who have collaborated are much more likely to be in the same partition than those who haven't. We can do the same thing for the Louvain clusters by running the following code:

```
plt.style.use('fivethirtyeight')
fig, axs = plt.subplots(1, 2, figsize=(18, 7), sharey=True)
charts = [(1, "have collaborated"), (0, "haven't collaborated")]

for index, chart in enumerate(charts):
    label, title = chart
    filtered = training_data.filter(training_data["label"] == label)
    values = (filtered.withColumn('sameLouvain',
            F.when(F.col("sameLouvain") == 0, "False")
                            .otherwise("True"))
        .groupby("sameLouvain")
        .agg(F.count("label").alias("count"))
        .select("sameLouvain", "count")
        .toPandas())
    values.set_index("sameLouvain", drop=True, inplace=True)
    values.plot(kind="bar", ax=axs[index], legend=None,
            title=f"Authors who {title} (label={label})")
    axs[index].xaxis.set_label_text("Same Louvain")

plt.tight_layout()
plt.show()
```

We can see the results of running that code in Figure 8-16.

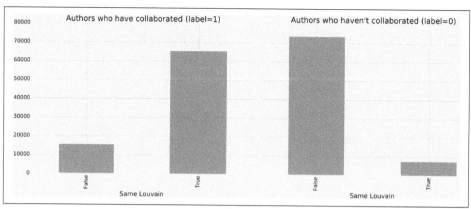

Figure 8-16. Same Louvain clusters

It looks like this feature could be quite predictive as well—authors who have collaborated are likely to be in the same cluster, and those who haven't are very unlikely to be in the same cluster.

We can train another model by running the following code:

```
fields = ["commonAuthors", "prefAttachment", "totalNeighbors",
          "minTriangles", "maxTriangles", "minCoefficient", "maxCoefficient",
          "samePartition", "sameLouvain"]
community_model = train_model(fields, training_data)
```

And now let's evaluate the model and display the results:

```
community_results = evaluate_model(community_model, test_data)
display_results(community_results)
```

The predictive measures for the community model are:

measure	score
accuracy	0.995771
recall	0.957088
precision	0.978674

Some of our measures have improved, so for comparison let's plot the ROC curve for all our models by running the following code:

```
plt, fig = create_roc_plot()

add_curve(plt, "Common Authors",
          basic_results["fpr"], basic_results["tpr"], basic_results["roc_auc"])

add_curve(plt, "Graphy",
          graphy_results["fpr"], graphy_results["tpr"],
          graphy_results["roc_auc"])
```

```
add_curve(plt, "Triangles",
          triangle_results["fpr"], triangle_results["tpr"],
          triangle_results["roc_auc"])

add_curve(plt, "Community",
          community_results["fpr"], community_results["tpr"],
          community_results["roc_auc"])

plt.legend(loc='lower right')
plt.show()
```

We can see the output in Figure 8-17.

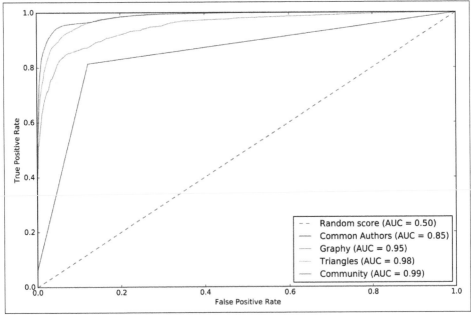

Figure 8-17. The ROC curve for the community model

We can see improvements with the addition of the community model, so let's see which are the most important features:

```
rf_model = community_model.stages[-1]
plot_feature_importance(fields, rf_model.featureImportances)
```

The results of running that function can be seen in Figure 8-18.

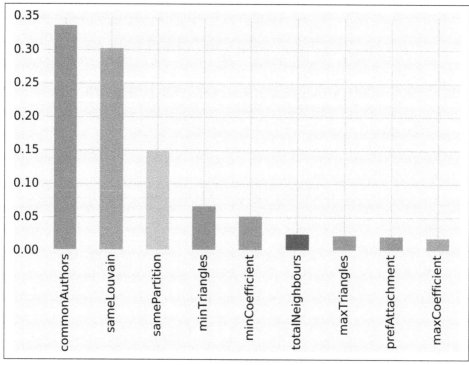

Figure 8-18. Feature importance: community model

Although the common authors model is overall very important, it's good to avoid having an overly dominant element that might skew predictions on new data. Community detection algorithms had a lot of influence in our last model with all the features included, and this helps round out our predictive approach.

We've seen in our examples that simple graph-based features are a good start, and then as we add more graphy and graph algorithm–based features, we continue to improve our predictive measures. We now have a good, balanced model for predicting coauthorship links.

Using graphs for connected feature extraction can significantly improve our predictions. The ideal graph features and algorithms vary depending on the attributes of the data, including the network domain and graph shape. We suggest first considering the predictive elements within your data and testing hypotheses with different types of connected features before fine-tuning.

> ### Reader Exercises
>
> There are several areas to investigate, and ways to build other models. Here are some ideas for further exploration:
>
> - How predictive is our model on conference data that we did not include?
> - When testing new data, what happens when we remove some features?
> - Does splitting the years differently for training and testing impact our predictions?
> - This dataset also has citations between papers; can we use that data to generate different features or predict future citations?

Summary

In this chapter, we looked at using graph features and algorithms to enhance machine learning. We covered a few preliminary concepts and then walked through a detailed example integrating Neo4j and Apache Spark for link prediction. We illustrated how to evaluate random forest classifier models and incorporate various types of connected features to improve our results.

Wrapping Things Up

In this book, we covered graph concepts as well as processing platforms and analytics. We then walked through many practical examples of how to use graph algorithms in Apache Spark and Neo4j. We finished with a look at how graphs enhance machine learning.

Graph algorithms are the powerhouse behind the analysis of real-world systems—from preventing fraud and optimizing call routing to predicting the spread of the flu. We hope you join us and develop your own unique solutions that take advantage of today's highly connected data.

Additional Information and Resources

In this section, we quickly cover additional information that may be helpful for some readers. We'll look at other types of algorithms, another way to import data into Neo4j, and another procedure library. There are also some resources for finding datasets, platform assistance, and training.

Other Algorithms

Many algorithms can be used with graph data. In this book, we've focused on those that are most representative of classic graph algorithms and those of most use to application developers. Some algorithms, such as coloring and heuristics, have been omitted because they are either of more interest in academic cases or can be easily derived.

Other algorithms, such as edge-based community detection, are interesting but have yet to be implemented in Neo4j or Apache Spark. We expect the list of graph algorithms used in both platforms to increase as the use of graph analytics grows.

There are also categories of algorithms that are used with graphs but aren't strictly graphy in nature. For example, we looked at a few algorithms used in the context of machine learning in Chapter 8. Another area of note is similarity algorithms, which are often applied to recommendations and link prediction. Similarity algorithms work out which nodes most resemble each other by using various methods to compare items like node attributes.

Neo4j Bulk Data Import and Yelp

Importing data into Neo4j with the Cypher query language uses a transactional approach. Figure A-1 illustrates a high-level overview of this process.

Figure A-1. Cypher-based import

While this method works well for incremental data loading or bulk loading of up to 10 million records, the Neo4j Import tool is a better choice when importing initial bulk datasets. This tool creates the store files directly, skipping the transaction log, as shown in Figure A-2.

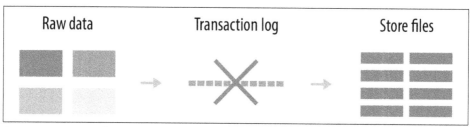

Figure A-2. Using the Neo4j Import tool

The Neo4j Import tool processes CSV files and expects these files to have specific headers. Figure A-3 shows an example of CSV files that can be processed by the tool.

	id:ID(User)	name	id:ID(Review)	text	stars
Nodes	1234	Bob	678	Awesome	3
	1235	Alice	679	Mediocre	2
	1236	Erika	680	Really bad	1

	:START_ID(User)	:END_ID(Review)
Relationships	1234	678
	1235	679
	1236	680

Figure A-3. Format of CSV files that Neo4j Import processes

The size of the Yelp dataset means the Neo4j Import tool is the best choice for getting the data into Neo4j. The data is in JSON format, so first we need to convert it into the format that the Neo4j Import tool expects. Figure A-4 shows an example of the JSON that we need to transform.

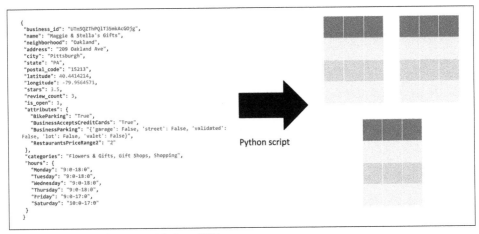

```
{
  "business_id": "UTmSQZThPQlT35mkAcGOjg",
  "name": "Maggie & Stella's Gifts",
  "neighborhood": "Oakland",
  "address": "209 Oakland Ave",
  "city": "Pittsburgh",
  "state": "PA",
  "postal_code": "15213",
  "latitude": 40.4414214,
  "longitude": -79.9564571,
  "stars": 3.5,
  "review_count": 3,
  "is_open": 1,
  "attributes": {
    "BikeParking": "True",
    "BusinessAcceptsCreditCards": "True",
    "BusinessParking": "{'garage': False, 'street': False, 'validated':
False, 'lot': False, 'valet': False}",
    "RestaurantsPriceRange2": "2"
  },
  "categories": "Flowers & Gifts, Gift Shops, Shopping",
  "hours": {
    "Monday": "9:0-18:0",
    "Tuesday": "9:0-18:0",
    "Wednesday": "9:0-18:0",
    "Thursday": "9:0-18:0",
    "Friday": "9:0-17:0",
    "Saturday": "10:0-17:0"
  }
}
```

Python script

Figure A-4. Transforming JSON to CSV

Using Python, we can create a simple script to convert the data to a CSV file. Once we've transformed the data into that format we can import it into Neo4j. Detailed instructions explaining how to do this are in the book's resources repository (*https:// bit.ly/2FPgGVV*).

APOC and Other Neo4j Tools

Awesome Procedures on Cypher (APOC) (*https://bit.ly/2JDfSbS*) is a library that contains more than 450 procedures and functions to help with common tasks such as data integration, data cleaning, and data conversion, and general help functions. APOC is the standard library for Neo4j.

Neo4j also has other tools that can be used in conjunction with their Graph Data Science library such as an algorithms "playground" app for code-free exploration. These can be found on their developer site for graph algorithms (*https://neo4j.com/devel oper/graph-algorithms*).

Finding Datasets

Finding a graphy dataset that aligns with testing goals or hypotheses can be challenging. In addition to reviewing research papers, consider exploring indexes for network datasets:

- The Stanford Network Analysis Project (SNAP) (*https://snap.stanford.edu/ index.html*) includes several datasets along with related papers and usage guides.

- The Colorado Index of Complex Networks (ICON) (*https://icon.colorado.edu/*) is a searchable index of research-quality network datasets from various domains of network science.

- The Koblenz Network Collection (KONECT) (*http://konect.uni-koblenz.de/*) includes large network datasets of various types in order to perform research in network science.

Most datasets require some massaging to transform them into a more useful format.

Assistance with the Apache Spark and Neo4j Platforms

There are many online resources for the Apache Spark and Neo4j platforms. If you have specific questions, we encourage you to reach out their respective communities:

- For general Spark questions, subscribe to *users@spark.apache.org* at the Spark Community page (*https://bit.ly/2UXMmyI*).

- For GraphFrames questions, use the GitHub issue tracker (*https://bit.ly/2YqnYrs*).

- For all Neo4j questions (including about graph algorithms), visit either the Neo4j documentation (*https://neo4j.com/docs/graph-data-science/current/algorithms/*) or the Neo4j Community online (*https://community.neo4j.com/*).

Training

There are a number of excellent resources for getting started with graph analytics. A search for courses or books on graph algorithms, network science, and analysis of networks will uncover many options. A few great examples for online learning include:

- Coursera's Applied Social Network Analysis in Python course (*https://bit.ly/2U87jtx*)

- Leonid Zhukov's Social Network Analysis YouTube series (*https://bit.ly/2Wq77n9*)

- Stanford's Analysis of Networks course (*http://web.stanford.edu/class/cs224w/*) includes video lectures, reading lists, and other resources

- Complexity Explorer (*https://www.complexityexplorer.org/*) offers online courses in complexity science

Index

About the Authors

Mark Needham is a graph advocate and developer relations engineer at Neo4j. He works to help users embrace graphs and Neo4j, building sophisticated solutions to challenging data problems. Mark has deep expertise in graph data, having previously helped to build Neo4j's Causal Clustering system. He writes about his experiences of being a graphista on his popular blog at *https://markhneedham.com/blog/* and tweets *@markhneedham* (*https://twitter.com/markhneedham*).

Amy E. Hodler is a network science devotee and AI and graph analytics program manager at Neo4j. She promotes the use of graph analytics to reveal structures within real-world networks and predict dynamic behavior. Amy helps teams apply novel approaches to generate new opportunities at companies such as EDS, Microsoft, Hewlett-Packard (HP), Hitachi IoT, and Cray Inc. Amy has a love for science and art with a fascination for complexity studies and graph theory. She tweets *@amyhodler* (*https://twitter.com/amyhodler*).

Colophon

The animal on the cover of *Graph Algorithms* is the European garden spider (*Araneus diadematus*), a common spider of Europe and also North America, where it was inadvertently introduced by European settlers.

The European garden spider is less than an inch long, and mottled brown with pale markings, a few of which on its back are arranged in such a way that they seem to form a small cross, giving the spider its common name of "cross spider." These spiders are common across their range and are most often noticed in late summer, as they grow to their largest size and begin spinning their webs.

European garden spiders are orb weavers, meaning that they spin a circular web in which they catch their small insect prey. The web is often consumed and respun at night to ensure and maintain its effectiveness. While the spider remains out of sight, one of its legs rests on a "signal line" connected to the web, movement on which alerts the spider to the presence of struggling prey. The spider then quickly moves to bite its prey to kill it and also inject it with special enzymes that enable consumption. When their webs are disturbed by predators or inadvertent disturbance, European garden spiders use their legs to shake their web, then drop to the ground on a thread of its silk. When danger passes, the spider uses this thread to reascend to its web.

They live for one year: after hatching in spring, the spiders mature during the summer and mate late in the year. Males approach females with caution, as females will sometimes kill and consume the males. After mating, the female spider weaves a dense silk cocoon for her eggs before dying in the fall.

Being quite common, and adapting well to human-disturbed habitats, these spiders are well studied. In 1973, two female garden spiders, named Arabella and Anita, were part of an experiment aboard NASA's *Skylab* orbiter, to test the effect of zero gravity on web construction. After an initial period of adapting to the weightless environment, Arabella built a partial web and then a fully formed circular web.

Many of the animals on O'Reilly covers are endangered; all of them are important to the world.

The cover image is a color illustration by Karen Montgomery, based on a black-and-white engraving from *Meyers Kleines Lexicon*. The cover fonts are Gilroy and Guardian Sans. The text font is Adobe Minion Pro; the heading font is Adobe Myriad Condensed; and the code font is Dalton Maag's Ubuntu Mono.

O'REILLY®

There's much more where this came from.

Experience books, videos, live online training courses, and more from O'Reilly and our 200+ partners—all in one place.

Learn more at oreilly.com/online-learning

9 781492 047681